# The Person Behind the Syndrome

Springer
London
Berlin
Heidelberg
New York
Barcelona
Budapest
Hong Kong
Milan
Paris
Santa Clara
Singapore
Tokyo

Peter Beighton · Greta Beighton

# The Person Behind the Syndrome

Foreword by Hans-R. Wiedemann

With 100 Illustrations

Springer

Peter Beighton, MD, PhD, FRCP, DCH
Greta Beighton, SRN, SCM, HV

Department of Human Genetics,
University of Cape Town Medical School,
Observatory 7925,
South Africa

*Cover:* Left to right in each row are: Sir William Richard Gowers; Angelo Maria Maffucci; Johann Friedrich Horner; William Anderson; Edward Nettleship; Willem Vrolik; William John Little; François de la Peyronie; Robert Marcus Gunn.

British Library Cataloguing in Publication Data
Beighton, Peter
  The person behind the syndrome
  1. Medical scientists  2. Syndromes  3. Diseases
  I. Title  II. Beighton, Greta
  610.9'22

Library of Congress Cataloging-in-Publication Data
Beighton, Peter.
  The person behind the syndrome/Peter Beighton & Greta Beighton.
      p. cm.
  Rev. ed. of: The man behind the syndrome/Peter Beighton, Greta
Beighton. c1986.
  Includes bibliographical references and index.
  ISBN-13: 978-1-4471-1236-5          e-ISBN-13: 978-1-4471-0925-9
  DOI: 10.1007/978-1-4471-0925-9
  1. Physicians – Biography.  2. Geneticists – Biography.  3. Medical
genetics.  I. Beighton, Greta, 1939–  .  II. Beighton, Peter. Man
behind the syndrome.  III. Title.
  [DNLM:  1. Physicians – biography.  2. Genetics, Medical – biography.
WZ 112 B417p 1996]
R134.B45 1996
610'.92'2–dc20
[B]
DNLM/DLC                                              96–17384
for Library of Congress                                    CIP



Typeset by EXPO Holdings, Malaysia

28/3830–543210 Printed on acid-free paper

*To our children, Charles, Victoria and Robert, now young adults, in the hope
that they will enjoy the precious gifts of good health, wisdom and good fortune*

# Foreword

A decade after the publication of *The Man Behind the Syndrome*, which was warmly received, particularly by medical geneticists, syndromologists and those doctors from many different disciplines with an interest in medical history, Peter and Greta Beighton now present the second volume of their work, promised ten years ago. The length of time which has passed since the publication of the first book gives an inkling of the extraordinary effort involved on the part of the authors in collecting the necessary biographical data and the portraits of their subjects.

*The Person Behind the Syndrome* conforms exactly in structure, quality and size with the first volume, thus facilitating the use of the series. Again we find detailed presentations of a hundred people who have given their names to disorders or syndromes which are thought to have a significant genetic or chromosomal component (with a photograph or portrait, biography, commentary on the development of nomenclature and references). The reader finds information not only on the doctor and/or scientist under discussion, but also, as in the previous volume, on the person behind the name. This is followed by brief, unillustrated biographies of about seventy, mostly younger and, in some cases, still professionally active personalities. Peter and Greta Beighton have made valiant efforts, as with their first volume, to include eponymous women but there are – still – very few of these and some could not be documented; however, whereas the first volume contained detailed or short biographies of only seven women, this new volume does at least record eleven.

Among the one hundred personalities presented here we find most of those eponymous authors whom we missed in the first volume, e.g., Fr. J. Kallmann (1897–1965) or Otto Ullrich (1894–1957); in addition, a subjectively chosen selection of twelve of the short biographies which appeared in 1986 have been expanded and included, with portraits, in the first section of this new book. The places of origin of these eponymous people are also interesting: the great majority are from Europe (122), followed by North America (45). If we compare the different fields of study of the personalities depicted in both books (although it is often difficult to slot them into neat categories), it is no surprise that the number of general practitioners as well as of pathologists is clearly falling while that of the representatives of genetics and of some clinical disciplines has increased.

The Beightons' work should not be understood as an attempt to build a pantheon for pioneers of syndromology; this would be not only very immodest, but also a false impression. With their excellent volumes *The Man Behind the Syndrome* (1986) and *The Person Behind the Syndrome* (1996) Peter and Greta Beighton have created a work which will please, inspire and motivate the historical and human interest of doctors and scientists, especially those working in the fields of clinical, medical and human genetics. These books will doubtless find many more new friends. May their authors amply harvest their well-earned thanks!

Kiel, Germany
January, 1996

Hans-Rudolf Wiedemann, Prof. Dr. med habil., Drs. h.c.

# Preface

In 1986, our curiosity concerning individuals whose names are attached to genetic disorders led to the compilation and publication of *The Man Behind the Syndrome*. We derived enormous enjoyment and satisfaction from that endeavour, together with considerable insight into human nature. In many instances, however, it was not possible to obtain portraits or biodata and there were some notable omissions from the book. Our investigations continued and as further material became available, we put together a new volume, choosing the title *The Person Behind the Syndrome* in deference to non-sexist principles. As in the previous book, we have presented portraits, biographical data, accounts of the evolution of the nomenclature and relevant references for 100 eponymous persons who are dead or long retired. In the second section, we have provided brief biographical sketches of a further 70 colleagues, the majority of whom are still active in their careers. We have included the latter individuals as it seems likely that we may not be in a position to write about them, when they have moved to the former category.

Cross references are provided in the text, especially where compound eponyms are concerned. Equally, if relevant material appears in *The Man Behind the Syndrome*, this is indicated. For the sake of clarity (and in the hope of stimulating sales) we have listed the eponyms which appeared in *The Man Behind the Syndrome* in the appendix.

Many of our friends have commented on the paucity of eponymous females in *The Man Behind the Syndrome*. We have gone to great lengths in our attempts to redress this imbalance in *The Person Behind the Syndrome* but, in reality, there are relatively few eponymous women. The reason is not chauvinism, simply that at the time that the major eponymic syndromes were established, only a small number of women had entered medicine. Indeed, for the first section of this book we were able to find portraits and biodata concerning only 9 women whose names are generally accepted as syndromic designations. The increase in the number of women now graduating in medicine is reflected to some extent in recent eponyms and we identified 9 women who warranted inclusion in the second section. Sadly, we only succeeded in contacting or eliciting a response from 2 of them, so females are still under-represented.

Chance has played a major role in eponymy, and many of the individuals whose names are well-known in the world of medical genetics made only minor contributions to the understanding of the condition which bears their eponym. On the other hand, there are those whose names amply warrant perpetuation but due to some quirk of circumstance, they escaped eponymy and eventually sank into obscurity. In recent times, there has been a tendency towards promiscuous eponymy; not infrequently, there are more names in a conjoined eponym than there are affected persons with the condition in question. In this book, we have included only those eponyms which are comparatively well known, or where the contribution has been significant. [For those who aspire to eponymous immortality, the trick is to identify a "new" syndrome and then to use an extremely cumbersome descriptive title in the initial report. If co-authorship can be avoided, so much the better. A single eponym, particularly if it is harmonious or curious, will stand a better chance of being perpetuated in further publications than a long obscure title!]

Cape Town                          Peter Beighton
January 1996                     Greta Beighton

# Acknowledgements

**Major contributions to our book, for which we are extremely grateful, were made by:**

Professor H.-R. Wiedemann of Germany
The late Professor David Klein of Switzerland (1908–1993)

**We offer special thanks for their assistance to:**

Professor T. Gedde-Dahl of Norway
Professor Robert Gorlin of Minneapolis, USA
Professor Judith Hall of Canada
Professor Norio Niikawa of Japan
Professor Pierre Maroteaux of Paris
Professor John Opitz of Montana, USA
Dr Anne De Paepe of Belgium
Professor H. Plauchu of Lyon, France
Professor Ilkka Kaitila of Finland
Dr Anton Brøgger of Norway
Dr Alex Paton of the British Postgraduate Medical Federation, London
Miss S. Katcher and the staff of the Medical Library, University of Cape Town

**We are grateful to many other librarians and medical historians for provision of biographical material and photographs:**

Lucinda Keister of the National Library of Medicine (History of Medicine Division), Bethesda, USA
Tina Cunnane, Nighat Ispahany and Anne M Pascarelli of the New York Academy of Medicine Library
I.F. Lyle, Librarian, Royal College of Surgeons of England
D. Stewart, Librarian, Royal Society of Medicine, London
G. Davenport, Librarian, Royal College of Physicians, London
Ms Barbara Watkins, Welch Medical Library, Johns Hopkins University

**We are appreciative of help which we received from friends and colleagues in the following fields:**

*Translations:*
Dr Jacquie Greenberg, Pierre Jansens, Jurgen Herre, Maja McCormack, Elmar Thomas and the late W. Heitner

*Preparation of the photographs:*
The staff of the Department of Medical Graphics, Groote Schuur Hospital

*Preparation of the manuscript and proof reading:*
Mrs Gillian Shapley

*Financial support for background research:*
University of Cape Town Staff Research Fund, South African Medical Research Council, Harry Crossley Foundation and the Mauerberger Fund

# Contents



# Section I. Portraits and Biographies

## Section II. Brief Biographies

# Bibliography


Ashwal S (ed) (1990) The founders of child neurology. Norman Publishing in association with the Child Neurology Society, San Francisco

Beighton P (ed) (1993) McKusick's heritable disorders of connective tissue, 5th ed. Mosby Baltimore, London, Sydney & Toronto

Bibliography of the history of medicine, National Library of Medicine (1964-1983) US Department of Health and Human Services, Public Health Service, National Institutes of Health, Bethesda, Maryland

Birch C (1979) Names we remember. Ravenswood, Beckenham, Kent, England

Brown GH (1955) Munks Roll. Vol IV. Lives of the Fellows of the Royal College of Physicians of London. The College, London

Clifford R, Bynum WF (1982) Historical aspects of the neurosciences. Raven, New York

Duke-Elder, Sir S (1962-1967) System of ophthalmology. Henry Kimpton, London

Garrison FH, Morton LT (1983) A medical bibliography. Gower, Aldershot, England

Hamilton Bailey, Bishop WJ (1959) Notable names in medicine and surgery. HK Lewis, London

Index Medicus, National Library of Medicine, US Department of Health and Human Services, Public Health Service, National Institutes of Health, Bethesda, Maryland

Jablonski S (1991) Dictionary of syndromes and eponymic diseases, 2nd ed. Krieger Publishing Company, Malabar, Florida

Jones KL (ed) (1988) Smith's recognizable patterns of human malformation, 4th edn. WB Saunders Co, Philadelphia & London

Kelly EC (1948) Encyclopaedia of medical sources. Williams & Wilkins, Baltimore, Maryland

Lebensohn JE (1969) An anthology of ophthalmic classics. Williams & Wilkins, Baltimore, Maryland.

Lindeboom GA (1984) Dutch medical biography. A biographical dictionary of Dutch physicians and surgeons 1475-1975. Rodopi, Amsterdam

Lourie J (1982) Medical eponyms: who was Condé? Pitman, London

Major RH (1945) Classical descriptions of disease. Charles C Thomas, Springfield, Illinois

McKusick VA (1994) Mendelian inheritance in man, 11th edn. The Johns Hopkins University Press, Baltimore & London

Munk W (1701-1825) Munk's roll. Vols II and III. The roll of the Royal College of Physicians of London. The College, Pall Mall East, London

Power Sir D'Arcy (1930) Plarr's lives of the Fellows of the Royal College of Surgeons of England. John Wright, Bristol, Simkin Marshall, London

Rang M (1966) Anthology of orthopaedics. Churchill Livingstone, Edinburgh, London & New York

Rett A (1990) Kinder in unserer Hand Ein Leben mit Behinderten. Orac Verlag, Wien, Frankfurt & Bern

Shelley WB, Crissey JT (1953) Classics in clinical dermatology. Charles C Thomas, Springfield, Illinois

Talbott JH (1970) A biographical history of medicine. Grune & Stratton, New York & London

Trail RR (1968) Munk's roll. Vol V. Lives of the Fellows of the Royal College of Physicians of London, The College, London

Webb Haymaker (ed) (1953) Founders of Neurology, 1st edn. Charles C Thomas, Springfield, Illinois

Wolstenholme G (1982) Munk's roll. Vol VI. Lives of the Fellows of the Royal College of Physicians of London. IRL Press Ltd, Oxford & Washington

# Section I

# Portraits and Biographies

# ADDISON, Thomas

*(1793–1860)*

A DDISON disease or adrenal insufficiency is usually acquired, but genetic forms have been recognised. Hyperpigmentation and hypoglycaemia are the major manifestations.

# Biography

A DDISON was a physician at Guy's Hospital, London, during the first half of the nineteenth century. He was a major figure in British medicine and is regarded as the founder of endocrinology.

Thomas Addison was born in 1793 at Long Benton, Northumberland, where his father was a grocer. He attended Newcastle Grammar School and qualified in medicine at the University of Edinburgh in 1815; he was a brilliant Latin scholar and took his lecture notes in that language. Addison developed an early interest in dermatology, as the title of his doctoral dissertation *De Syphilide et Hydrargyro* shows.

After an appointment as house surgeon at the Lock Hospital, Addison became a "perpetual physician's pupil" at Guy's Hospital, London, where he spent the rest of his career. He rose steadily through the hierarchy at Guy's; in 1824 he was appointed as assistant physician and in 1837 he became lecturer in medicine and full physician in conjunction with the great Richard Bright. In the following year Addison was elected to the Fellowship of the Royal College of Physicians. Other luminaries at Guy's during this period included Gull, Wilkes and Hodgkin; this formidable group brought immense prestige to their hospital.

Addison founded the Department of Dermatology at Guy's in 1824 and his influence is still evident in the collection of wax models of skin disorders which were prepared under his supervision. His achievements in dermatology included classical accounts of xanthoma diabeticorum, morphea and keloid. He was also interested in chest disease, and made notable contributions towards the understanding of the pathogenesis of pulmonary infection. As a general physician, Addison gave his name to pernicious or Addisonian anaemia, and to adrenocortical insufficiency, or Addison disease (vide infra). He introduced the use of static electricity in the treatment of spasticity and he co-authored books on toxicology and general medicine; the latter contains the first clear description of appendicitis.

Addison was totally committed to Guy's Hospital and he had few interests outside his work. He was regarded as being difficult, haughty and distant, but his great clinical skills and lecturing abilities earned him the respect of his students and colleagues. Nevertheless, it is said that he was happier in the post-mortem room than in the wards.

Addison suffered from severe bouts of depression, and contemplated suicide on several occasions. In 1845, at the age of 52 years, Addison married a widow with 2 children; thereafter his depression worsened and in March 1860 he retired from practice and wrote to his students in the following terms: "A considerable breakdown in my health has scared me from the anxieties, responsibilities and excitement of the profession; whether temporarily or permanently cannot yet be determined but, whatever may be the issue, be assured that nothing was better calculated to soothe me than the kind interest manifested by the pupils of Guy's Hospital during the many trying years devoted to that institution."

Three months later he died from injuries sustained when he threw himself out of a window in Brighton.

Addison was buried in Lanercost Abbey, Cumberland, near his childhood home.

# Nomenclature

T HE adrenal glands were described by Eustachius in 1714 but it was many years before their function was elucidated. Indeed, the offer of a substantial prize for an essay on adrenal physiology, made by the Académie des Sciences de Bordeaux in the eighteenth century failed to elicit any significant entries.

In 1849 Addison presented a paper to the South London Medical Society entitled "*Anaemia-Disease of the Supra-renal Capsules*. This presentation contained the first clear account of pernicious anaemia, to which his name is now attached (Addisonian anaemia), and it also demonstrated his interest in the adrenals. Addison consolidated his knowledge of these organs in an illustrated monograph published in 1855, entitled *On the Constitutional and Local Effects of Disease of the Supra-renal Capsules*. Addison's detailed description of the manifestations of adrenal cortical insufficiency, which appeared in this work, ensured his place as the founder of the science of endocrinology.

The term "Addison's disease" came into general use for adrenal insufficiency, from whatever cause. Tuberculosis was regarded as the most common pathogenic factor, but it subsequently became apparent that the adrenals could be damaged by other mechanisms. With the advent of modern endocrinology and the wider understanding of genetic mechanisms, familial forms of Addison's disease were recognised. These include autosomal recessive adrenal unresponsiveness to ACTH, or familial glucocorticoid deficiency, which was initially termed "familial Addison's disease" by Shepard et al. (1959). In another autosomal recessive disorder, Addison disease is associated with diabetes mellitus and myxoedema (Phair et al. 1965); this condition is also known as the Schmidt syndrome or polyglandular autoimmune syndrome type II. An X-linked combination of Addison disease and cerebral sclerosis is well documented (Fanconi et al. 1964); this condition is now known as adrenoleukodystrophy or adrenomyeloneuropathy. A further X-linked form of adrenal hypoplasia, with specific histological changes has also been recognised (Martin 1971). Further heterogeneity is likely.

# References


Addison T (1849) Anaemia: Disease of the supra-renal capsules. London Hosp Gaz 43: 517–518

Fanconi A, Prader A, Isler W, Luthy F, Siebenmann RE (1964) Morbus Addison mit Hirnsklerose im Kindesalter. Ein hereditaeres Syndrom mit X-chromosomaler Vererbung? Helv Paediatr Acta 18: 480–501

Martin MM (1971) Familial Addison's disease. Birth Defects Original Article Series VII(6): 98–100

Phair JP, Bondy PK, Abelson DM (1965) Diabetes mellitus, Addison's disease and myxoedema: report of two cases. J Clin Endocr 25: 260–265

Shepard TH, Landing BH, Mason DG (1959) Familial Addison's disease. Case reports of two sisters with corticoid deficiency unassociated with hypoaldosteronism. Am J Dis Child 97: 154–162

Sutherland FM (1960) Nova et Vetera. Thomas Addison 1793–1860. Br Med J 5194: 304–305

# ADIE, William John

*(1886–1935)*

*From:* Egerton Caughey J. (1953) In: Webb Haymaker (ed) The
Founders of Neurology, 1st edn

*Courtesy:* Charles C. Thomas, Springfield, Illinois
Dr Macdonald Critchley, London, England

$\mathbf{A}$ DIE-HOLMES syndrome is a benign disorder in which dilated, unresponsive pupils are associated with absence of the tendon reflexes. Inheritance is probably autosomal dominant.

# Biography

$\mathbf{A}$ DIE was a distinguished academic neurologist in London during the earlier years of the twentieth century.

William Adie was born in Geelong, Victoria, Australia on 31 October 1886. He received his education in that town at the Flinders School and then studied medicine at the University of Edinburgh, qualifying in 1911. The award of a postgraduate travelling scholarship enabled him to gain experience in Munich, Berlin, Vienna and Paris, prior to taking up an appointment as resident medical officer at the National Hospital for the Paralysed and Epileptic, London.

Adie joined the Northamptonshire Regiment as medical officer at the outbreak of World War I, and was amongst the survivors of the retreat from Mons. He was then transferred to the Leicestershire Regiment and saw extensive active service during the next 2 years. In April 1916 Adie was mentioned in despatches for gallantry in the field during an enemy gas attack. He subsequently took charge of the 7th General Hospital, and also acted as a consultant in the management of head injuries.

After the armistice, Adie became medical registrar at the Charing Cross Hospital, London. He obtained the Fellowship of the Royal College of Physicians in 1925 and in this same year received a gold medal from the University of Edinburgh for his doctoral thesis on idiopathic narcolepsy. His career culminated with appointments as consultant neurologist at the Charing Cross Hospital and the National Hospital, Queen's Square.

Adie contributed extensively to the neurological literature and the subjects which he addressed included migraine, pituitary tumours and disseminated sclerosis. In conjunction with his friend and mentor, James Collier, Adie was also joint editor of the neurological section of Price's classical *Textbook of Medicine*.

Adie had great intellectual gifts and he was an acute clinical observer and a fine teacher. He was also a kindly, modest, approachable man and he was held in high regard by his students and colleagues. In 1916 Adie married Lorraine Bonar and the couple had a son and daughter. He lived an active life and had many interests. In particular, he was a keen ornithologist and tennis player and he enjoyed skiing and skating during his holidays in Switzerland.

Adie had a coronary thrombosis in 1932 and for the next three years he suffered considerable ill health. His career as clinician, teacher and medical author came to an end when he died of a further myocardial infarction on 17 February 1935.

# Nomenclature

$\mathbf{I}$ N 1931, while associated with the Moorfields Ophthalmological Hospital, Adie published an account of pupils which reacted to accommodation but not to light, and which were associated with depressed or absent tendon reflexes. In order to draw distinction from the similar pupillary features produced by the effects of syphilis on the central nervous system, Adie used the term "pseudo Argyll Robertson pupils". Gordon Holmes published concurrently on the same subject using the term "partial iridoplegia" (*see* p. 83).

In 1932 Adie wrote a further paper on the disorder, drawing attention to the benign nature of the condition and emphasising the clinical variability and the absence of tendon reflexes. Thereafter the condition was known as the Adie syndrome or Adie pupil. The conjoined eponym "Holmes-Adie" also came into use at this time. A full description of the disorder had previously been published in the French literature by Weill and Reys (1926) while Strassburger (1902) and Saenger (1902) had documented the pupillary features without mentioning the tendon areflexia. In the past these eponyms were occasionally added to that of Adie but they are rarely mentioned in the modern literature.

The Adie syndrome has been documented in successive generations and it seems likely that it is the result of genetically determined degeneration of the ciliary and dorsal root ganglia. Although listed in McKusick's catalogue *Mendelian Inheritance in Man*, the Adie syndrome has not yet been accorded the asterisk of syndromic identity.

In addition to his publications on pupillary abnormalities, Adie's work on narcolepsy attracted wide attention when it was published in 1926. Several other authors had previously written comprehensive accounts of this disorder, but the eponym "maladie d'Adie" is still sometimes used in the French literature.

# References


Adie WJ (1931) Pseudo-Argyll Robertson pupils with absent tendon reflexes. A benign disorder simulating tabes dorsalis. Br Med J 1: 928–930

Adie WJ (1932) Tonic pupils and absent tendon reflexes: A benign disorder sui generis: Its complete and incomplete forms. Brain 55: 98–113

Holmes G (1931) Partial iridoplegia associated with symptoms of other disease of the nervous system. Tr Ophth Soc UK 51: 209–228

Obituary – William John Adie (1935) J Nerv Ment Dis 81: 726

Saenger A (1902) Myotonische Pupillenbewegug. Zbl Neurol 21: 837–839

Strassburger (1902) Zbl Neurol 21: 837–839

Weill G, Reys L (1926) Sur la pupillatonie. Contribution à l'étude de sa pathogénie. A propos d'un cas de réaction tonique d'une pupille à la convergence et parèsie de l'accomodation avec aréflexie à la lumière chez un sujet atteint de crises tétaniformes et d'aréflexie des membres inférieurs. Rev Otoneurocul Paris 4: 433–441

# ALDRICH, Robert A.
*(b. 1917)*

W ISKOTT-ALDRICH syndrome comprises eczema, thrombocytopenia and frequent infections due to immunological deficiency. Inheritance is X-linked recessive.

# Biography

A LDRICH had a distinguished career in academic paediatrics in the USA and made major contributions to the understanding of human development.

Robert Aldrich was born on 13 December 1917, in Evanston, Illinois, where his father was a paediatrician. His mother was a gifted writer and his parents collaborated on a classic book entitled *Babies are Human Beings*. Aldrich had a happy childhood and enjoyed contact with numerous international experts in childhood development, who visited his home from all parts of the world. Not surprisingly, this background had a profound influence on the direction of Aldrich's future career.

Aldrich was educated at Amherst College, Massachusetts and at the Northwestern University Medical School, Chicago, where he obtained his MD with highest distinction in 1944. During the next two years he had extensive service with the US Navy as a medical officer in the South Pacific. He then trained in paediatrics at the University of Minnesota, and in 1950 he joined the consultant staff of the Mayo Clinic. He moved to an academic post at the University of Oregon in 1951 and five years later he was elevated to the Chair of Paediatrics at the University of Washington in Seattle. In 1962 Aldrich accepted an invitation from President John F. Kennedy to take leave of absence and form a new National Institute of Child Health and Human Development at the NIH, Washington, DC. After successfully completing this task Aldrich returned to his own university and in 1964–5 he founded the Division of Human Ecology. In 1970 he became vice-president for Health Affairs at the University of Colorado, and occupied this post for the next decade. Aldrich returned to the University of Washington in 1980 and since that date he has been involved in a project entitled "The Child in the City", which seeks to make the metropolitan area an excellent place in which to rear children.

In addition to his activities in clinical and developmental paediatrics, Aldrich has had formal training in biochemistry and he has also occupied the Chair of Anthropology at the University of Colorado. His interests extend to city planning, which is interlinked with his child development project.

In 1989 Aldrich was president of the International Association for Child Ecology. In this context, he summed up his personal philosophy that "human development is peace. The question about whether we are capable of human development in the fullest sense remains unanswered".

# Nomenclature

A LDRICH gave the following personal account of the way in which the syndrome bearing his name was delineated. "The circumstances that led to describing the genetics of the Wiskott-Aldrich syndrome are of clinical interest. I was at the time a full-time member of the Mayo Clinic consultant staff. In this role in the Section of Pediatrics, each of us took a turn at attending on the inpatient service at St Mary's Hospital in Rochester, Minnesota. I was on that service when the little boy who was the index case was admitted in serious condition, as described in our published article. The clinical picture was one that I had never seen or read about in the pediatric literature. Neither had any other members of our full-time staff. Puzzled and looking for etiologic clues that might come from the child's environment or the history of the family, I invited the mother to sit down with me once more and go into more detail about the clinical course and possible cause of his serious illness. When she arrived, her mother accompanied her for the first time. After at least an hour of questioning about the environment that failed to add any leads, I began asking about relatives who might have had a similar illness. It was then that the child's grandmother exclaimed sadly, 'Just like all the rest of them'. This led to a search for each male death in the family and to establish the sex-linked nature of the syndrome. We were able to trace the female carrier, who came from the Netherlands to live in Iowa. This fascinating step was made possible through the co-operation of pediatricians in the area of the Netherlands from which she came. It is important to mention that we did not publish for more than a year after the research study was completed. I could not believe that this pattern had not been described. The literature (English and foreign) was fine combed by the Mayo Clinic staff without finding any reference to a syndrome resembling this one. Finally, together with my colleagues, Arthur Steinberg and Donald Campbell, an account was published in *Pediatrics* using the title *Pedigree demonstrating a sex-linked recessive condition characterized by draining ears, eczematoid dermatitis and bloody diarrhea*. It was several months later that a prominent German pediatrician wrote, directing my attention to a short abstract in a journal by Wiskott (1937) describing the clinical picture precisely in 3 brothers. Our contribution, of course, is the study of the genetic inheritance. That's the brief story. It illustrates the importance of a complete history."

Since Aldrich's original paper appeared in 1954, there have been more than 40 publications concerning the syndrome and the conjoined eponym is well established. In a multi-author review, Perry et al. (1980) estimated that the condition was present in 4.0 per million live births in the USA. Fischer et al. (1986) were able to assemble and analyse a series of 162 affected persons from European centres in whom therapeutic bone marrow transplantation had been undertaken. Current interest centres around the molecular genetics of the syndrome (*see The Man Behind the Syndrome* Wiskott, p. 231).

# References

Aldrich RA, Steinberg AG, Campbell DC (1954) Pedigree demonstrating a sex-linked recessive condition characterized by draining ears, eczematoid dermatitis and bloody diarrhea. Pediatrics 13(2): 133–139

Fischer A, Friedrich W, Levinsky R, Vossen J, Criscelli C, Kubanek B, Morgan G, Wagemaker G, Landais P (1986) Bone marrow transplantation for immunodeficiencies and osteopetrosis: European survey, 1968–1985. Lancet II: 1080–1084

Perry GS III, Spector BD, Schuman LM, Mandel JS, Anderson VE, McHugh RB, Hanson MR, Fahlstrom SM, Krivit W, Kersey JH (1980) The Wiskott-Aldrich syndrome in the United States and Canada (1892–1979). J Pediat 97: 72–78

Wiskott A (1937) Familiaerer, angeborener Morbus Werlhofii? Monatsschr Kinderheilk 68: 212–216

# ALPERS, Bernard J.

*(1900–1981)*

*From:* Postgraduate Medicine (1950) 7: 417

*Courtesy:* Postgraduate Medicine, Minneapolis, USA

$A$LPERS progressive infantile poliodystrophy is a lethal autosomal recessive neurodegenerative disorder in which the grey matter of the brain is predominantly involved. Epilepsy, failure to thrive, myoclonus, hypotonia and spasticity are major features and death in infancy is usual.

# Biography

$A$LPERS was a distinguished North American neurologist during the middle period of the twentieth century.

Bernard J. Alpers was born in Salem, Massachusetts, in 1900 and educated at Harvard College and Harvard Medical School. After obtaining his medical qualification, Alpers undertook advanced training in neuropathology in Europe, before taking up an appointment in the Department of Neurosurgery at the University of Pennsylvania. In 1939 he became head of the Department of Nervous and Mental Disease at Jefferson Medical College, Philadelphia. He soon established an independent Department of Neurology, where he was chairman for the next 25 years. During his career Alpers served as president of the American Association of Neuropathology, the American Neurological Association and the American Board of Psychiatry and Neurology.

Alpers was an excellent teacher with a keen sense of humour and his grand rounds attracted large audiences. He had highly developed clinical skills and his opinion on difficult neurological problems was greatly valued. Alpers commenced his ward rounds at Jefferson at 6.00 a.m. in order to demonstrate that there was a need for the maximum amount of time to do what was required in the course of the day. Although these legendary pre-dawn rounds were regarded with approval at the time, it is questionable whether they would be acceptable to patients and staff in the present era! Cynics might suggest that Alpers lacked the ability to organise his time efficiently!

Alpers published extensively and in 1941 he wrote his famous textbook *Clinical Neurology* which went into 5 editions; he was also editor of successive editions of *Essentials of Neurological Examination*. His many academic contributions were recognised by the award of the degree of doctor of science in medicine.

Alpers had a large personal library; he was a Hebrew scholar and he had a life-long interest in religion. Music was a constant source of pleasure to him and, at a different level, he followed the fortunes of the Boston Red Sox baseball team.

Alpers became emeritus professor of neurology at the Thomas Jefferson University on his retirement in 1965.

# Nomenclature

$I$N 1931, during the early phase of his career, Alpers gave a detailed description of a neurodegenerative disorder which was characterised by diffuse, progressive degeneration of the grey matter of the cerebrum. The disorder was differentiated into infantile and juvenile types on a basis of natural history (Ford et al. 1951). Alpers' eponym was employed by Blackwood et al. (1963) and the descriptive term "poliodystrophia cerebri progressiva infantilis" was introduced by Christensen and Hojgaard (1964) in a description of affected siblings.

The phenotype was expanded by Wefring and Lamvik (1967) when they drew attention to hepatic involvement in an affected brother and sister. Thereafter the cumbersome term "Alpers diffuse degeneration of cerebral grey matter with hepatic fibrosis" gained some favour, but the short eponymous form is now in general use. With advances in the understanding of the biochemical basis of the disorder, it is becoming increasingly evident that Alpers disease is a nonspecific heterogeneous group of disorders.

# References


Alpers BJ (1931) Diffuse progressive degeneration of the gray matter of the cerebrum. Arch Neurol Psychiat 25: 469–505

Blackwood W, Buxton PH, Cumings JN, Robertson DJ, Tucker SM (1963) Diffuse cerebral degeneration in infancy (Alpers' disease). Arch Dis Child 38: 193–204

Christensen E, Hojgaard K (1964) Poliodystrophia cerebri progressiva infantilis. Acta Neurol Scand 40: 21–40

Ford FR, Livingston S, Pryles CV (1951) Familial degeneration of the cerebral gray matter in childhood with convulsions, myoclonus, spasticity, cerebral ataxia, choreoathetosis, dementia and death in status epilepticus. Differentiation of infantile and juvenile types. J Pediat 39: 33–43

Wefring KW, Lamvik JO (1967) Familial progressive poliodystrophy with cirrhosis of the liver. Acta Paediatr Scand 56: 295–300

# ANDERSON, William

## *(1842–1900)*

*From:* St Thomas's Hosp Rep (1901) 30: 329

*Courtesy:* United Medical and Dental Schools of Guy's and St
Thomas's Hospitals Library Services, University of London

A NDERSON-FABRY disease, or angiokeratoma corporis diffusum universale, is an X-linked disorder. The main manifestations are burning sensations in the hands and feet, diffuse dark nodular skin lesions and progressive renal failure.

# Biography

W ILLIAM ANDERSON was a distinguished anatomist and surgeon at St Thomas's Hospital, London, during the latter decades of the nineteenth century. He had close links with Japan and was an expert in oriental art.

William Anderson was born in London on 18 December 1842 and educated at the City of London School. He then entered the University of Aberdeen but soon moved to the Lambeth School of Art, where he developed his considerable talents as a draftsman and artist.

Anderson became a medical student at St Thomas's Hospital in 1864 and received several medals and awards before qualifying in 1867. He was a hard-working student and, being somewhat reticent, preferred the wards and class rooms to boisterous extra-curricular activities. Anderson obtained the Fellowship of the Royal College of Surgeons in 1869 and thereafter gained practical surgical experience at the General Hospital, Derby, before returning to St Thomas's in 1871. His appointment as surgical registrar and demonstrator of anatomy coincided with the opening of the new hospital, which was situated on the south bank of the River Thames opposite the Houses of Parliament. His artistic abilities proved to be of great value in the illustration of his lectures on anatomy and he quickly acquired a reputation for his prowess in this field. His students were especially impressed by his ability to draw on the blackboard simultaneously with both hands!

In 1873 Anderson married the daughter of Dr Hall, a physician of Derby. He then sailed for Japan with his newly wedded wife in response to an invitation from the Japanese government to take up an appointment as Director of the Naval Medical College, Tokyo. Anderson remained in Japan for 7 years, lecturing on anatomy, physiology and surgery and living in the English colony, where he acted as medical officer to the British Legation. He had a genial nature and his popularity with his students was enhanced by the fluency which he developed in the Japanese language.

Anderson's artistic interests flourished during his stay in Tokyo and he became an avid collector of Japanese pictures. His first collection was destroyed in a fire in which he lost most of his possessions, but he was soon able to replace his losses. He eventually put together a series of drawings which illustrated the history and development of Japanese art. This collection was purchased by the British Museum, where it was regarded as having great historical importance. In his later years Anderson wrote authoritative texts on the subject of oriental art and he maintained his links with the Japanese community in London. His contributions were eventually recognised by the bestowal of the Order of the Rising Sun, with rank of commander.

In 1880 Anderson returned to St Thomas's as assistant surgeon and lecturer in anatomy, in place of W.W. Wagstaffe, who had retired due to ill health (paradoxically, Wagstaffe survived to write Anderson's obituary, 20 years later!). Thereafter he climbed steadily up the academic ladder, taking over the directorship of the department for diseases of the skin in 1887 and becoming full surgeon in 1891. Surgery and dermatology seem to be an unlikely combination but Anderson's artistic eye and appreciation of shape, colour and texture probably attracted him to the latter speciality.

Anderson had a special interest in the history of surgery and he produced a number of articles on this subject. In addition, he made academic contributions in the fields of general surgery, anatomy and dermatology. Once he had reached senior status Anderson was a regular examiner in Anatomy and Surgery for the Royal College and the University of London.

Anderson moved in medical and artistic circles, where he was liked and respected for his wide knowledge, polished manners, good nature and modest demeanour. In his professional life he was known for his diligence, integrity and conscientiousness. Anderson died suddenly on 27 October 1900 while driving through London in his carriage. Autopsy revealed rupture of a cord of the mitral valve.

# Nomenclature

I N 1898 Anderson published an account of a male aged 39 years who had an eruption on his trunk, genitals and proximal limbs. He recorded that the patient had been afflicted since childhood and that varicose veins, rectal bleeding and albuminuria had developed. Anderson termed the condition "angiokeratoma" and suggested that there might be generalised changes in the vascular system.

By one of those strange chance concurrences which happen in medical research, Fabry, a dermatologist in Dortmund, conducted independent studies of an affected boy and published his findings in the same year as Anderson. In his article Fabry used the designation "purpura haemorrhagica nodularis". A further case was recognised in Egypt by Madden (1912) and the condition was mentioned again by Fabry (1915) under the title "angiokeratoma corporis naeviforme".

Fabry retained his interest in the disorder and published the autopsy findings after his patient's death in 1930. He has been widely criticised for omitting any mention of Anderson and although the eponym "Fabry" is still used, it is being replaced by the more accurate conjoined form "Anderson-Fabry disease". Current interest centres around the elucidation of the basic defect and for this reason alternative designations such as hereditary dystopic lipidosis, alpha-galactosidase A deficiency and ceramide trihexosidase deficiency have found some favour (*see The Man Behind the Syndrome* Fabry, p. 52).

# References


Anderson W (1898) A case of "angio-keratoma". Br J Dermatol 10: 113–117

Fabry J (1898) Ein Beitrag zur Kenntnis der Purpura haemorrhagica nodularis (Purpura papulosa haemorrhagical Hebrae). Arch Dermatol Syphilol 43: 187–200

Fabry J (1915) Ueber einen Fall von Angiokeratoma circumscriptum am linken Oberschenkel. Dermatol Z 22: 1–4

Fabry J (1916) Zur Klinik und Aetiologie der Angiokeratoma. Arch Dermatol Syphilol 123: 294–307

Fabry J (1930) Weiterer Beitrag zur Klinik des Angiokeratoma naeviforme (Naevus angiokeratosus). Dermatol Wschr 90: 339–341

Madden FC (1912) Papilliform lesions (lymphangiomata) of the scrotum. Br Med J 2: 302–304

Payne JF (1901) William Anderson. St Thomas's Hospital Journal 30: 329–335

Wagstaffe WW (1901) A personal account of an old friend. St Thomas's Hospital Journal 30: 337–343

# ANGELMAN, Harry

*(b. 1915)*

A NGELMAN, or Happy Puppet syndrome, comprises mental retardation, microbrachycephaly and characteristic stiff, jerky, ataxic, puppet-like movements, a happy disposition and episodes of spontaneous laughter. The condition is the consequence of a defect in the long arm of chromosome 15.

# Biography

A NGELMAN is a British paediatrician, known for the disorder which bears his name.

HARRY ANGELMAN was born in 1915 in Birkenhead, England and qualified in medicine at the University of Liverpool in 1938. After junior hospital appointments he was commissioned into the Royal Army Medical Corps for service during World War II. Angelman spent several years with the military in India and after his return to civilian life in 1946 undertook postgraduate training in paediatrics. In 1950 Angelman was appointed as consultant paediatrician at Warrington General Hospital, Lancashire, UK, and he remained in this post until his retirement in 1976.

In 1995 Angelman was living in Lee-on-Solent on the south coast of England, with his wife Audrey Stuart, née Taylor, whom he had married in 1964. In his retirement his main interest has been the translation of Italian medical literature. Angelman and his wife have retained links with lay societies for families with Angelman syndrome in the UK, Europe and the USA.

# Nomenclature

I N 1965 Angelman published an account of the clinical features of children under his care, introducing his article as follows: "The association of mental retardation with abnormal physical development of congenital origin still includes a great variety of conditions whose causation is undetermined and which lack precise classification. The children described here possess such similarities as to justify combining them into a specific group, as yet of unknown cause. Their flat heads, jerky movements, protruding tongues and bouts of laughter give them a superficial resemblance to puppets, an unscientific name but one which may provide for easy identification."

The title of the article *Puppet Children* was derived from an oil painting by Caroto (1480–1555), entitled *Boy with a Puppet* which Angelman had observed in the Castelvecchio Museum in Verona, Italy. Angelman commented, "the laughing face of the boy in the painting and the association of the word 'puppet' with the puppet-like movements of these children crystalised the picture of these children in my mind, hence the title".

Two years after Angelman's description, Bower and Jeavons (1967) reported 2 affected children and modified the title to "Happy Puppet Syndrome". This term came into general use, appearing in the French language literature as "marionette joyeuse" or "syndrome du pantin hilare". Numerous reports followed and in 1982 Williams and Frias suggested that the eponym "Angelman" should replace the descriptive title of the condition, in order to avoid any possible offence to the families of affected persons.

Interest in the Angelman syndrome has continued and it has now been shown that a significant proportion of persons with the condition have a small deletion in the long arm of chromosome 15. This chromosomal site is very close to that which is involved in the Prader-Labhart-Willi syndrome, and it seems likely that the process of genomic imprinting is operative. A perspective of the frequency and medical importance of the condition can be gleaned from the fact that Angelman Syndrome Support groups are active in North America and Britain, having respective memberships of the families of about 800 and 200 affected children.

# References


Angelman H (1965) "Puppet" Children: a report on three cases. Develop Med Child Neurol 7: 681–688

Bower BD, Jeavons PM (1967) The "happy puppet" syndrome. Arch Dis Child 42: 298–301

Halal F, Chagnon J (1976) Le syndrome de la "marionette joyeuse". Union Med Can 105: 1077–1083

Knoll JHM, Nicholls RD, Magenis RE, Graham JM Jr, Lalande M, Latt SA (1989) Angelman and Prader-Willi syndromes share a common chromosome 15 deletion but differ in parental origin of the deletion. Am J Med Genet 32: 285–290

Pelc S, Levy J, Point G (1976) "Happy puppet" syndrome ou syndrome du "pantin hilare". Helv Paediatr Acta 31: 183–188

Williams CA, Frias JL (1982) The Angelman ("happy puppet") syndrome. Am J Med Genet 11: 453–460

# AXENFELD, Theodor
*(1867–1930)*

A XENFELD-RIEGER anomaly comprises defects of the anterior segment of the eye and iris, together with a characteristic facies and minor skeletal abnormalities. Familial aggregation has been documented but the mode of inheritance is uncertain.

# Biography

A XENFELD was professor of ophthalmology at the University of Freiburg, Germany, in the early decades of the twentieth century. He made many contributions to the understanding of the pathology of eye disease.

Theodor Axenfeld was born in 1867 in Smyrna where his father, a German, held an ecclesiastical post. The family returned to Godesberg, Germany, during his childhood and he entered the University of Bonn. Axenfeld then studied medicine at the Universities of Marburg and Berlin, qualifying in 1890. He undertook postgraduate training in ophthalmology under Uhthoff and, after developing an interest in the pathology of eye disease, spent a period in the laboratory of the great Von Helmholtz. In 1897, at the age of 30 years, Axenfeld accepted the chair of ophthalmology at the University of Rostock, and in 1901 he moved to a similar post at the University of Freiburg, where he remained until his death in 1930.

Axenfeld had a special interest in infections of the eye and in Freiburg his excellent clinical and laboratory facilities enabled him to expand his researches in this field. In 1902 he published a monograph on trachoma and this was followed by numerous articles on tuberculosis of the eye. By 1907 Axenfeld had accumulated sufficient experience to publish a book on ophthalmological bacteriology and in 1909 he published his *Textbook of Ophthalmology*. This book became the standard work in its field and by 1923 it had run to seven editions. In addition to his books Axenfeld was the author or co-author of almost 200 medical articles and he also served for 30 years as editor of the *Klinische Monatsblätter für Augenheilkunde*.

Axenfeld had great intellectual powers, combined with a critical mind, diligence and the ability to focus on the problem at hand. He was a humanitarian, with deep religious sentiments and an upright character. These academic and personal attributes gained for him the admiration and respect of his colleagues and friends.

The excellence of his ophthalmological service, his teaching abilities and his operative skills attracted many postgraduates and Axenfeld gained a secure international reputation. In 1925 he was elected to the presidency of the German Ophthalmological Society from whom he received the Graefe medal. He was also the recipient of the gold medal of the American Ophthalmological Society for his services to the speciality.

Axenfeld maintained his acute intellectual facilities until the end of his career. He became unwell after a brief academic visit to Japan and, following an abdominal operation in Freiburg, he died on 29 July 1930 at the age of 64 years.

# Nomenclature

T HE structural abnormalities of the anterior segment of the eye are a complex group of developmental defects which have been the subject of several overlapping classifications. The terms "posterior embryotoxon of Axenfeld" and "Axenfeld anomaly" pertain to enlargement and displacement of Schwalbe's line, with prominent iris processes in the anterior chamber and alterations in pupillary position, shape and number (Axenfeld 1920). The Axenfeld anomaly is usually mild but it is very variable in severity and occasionally complicated by glaucoma.

The name of Herwigh Rieger, 1898–1986 is linked to a disorder of mesodermal dysgenesis, which comprises the Axenfeld anomaly plus hypoplasia of the stroma of the iris. Pupillary abnormalities may be present and glaucoma is a recognised complication.

The separate identity of the Axenfeld and Rieger anomalies is uncertain, as is their genetic basis, but they occur as an uncommon autosomal dominant trait in combination with facial, dental and skeletal abnormalities. This condition, which is termed the Axenfeld-Rieger syndrome, was documented in 1973 in a mother and her three offspring by De Hauwere et al. and in 1991 Chitty et al. reported a further affected family. The syndromic boundaries of this disorder are not fully defined and syndromic status is not yet secure (*see The Man Behind the Syndrome* Rieger, p. 147).

# References


Axenfeld T (1920) Ber Deut Ophth Ges 43: 301

Chitty LS, McCrimmon R, Temple IK, Russell-Eggitt IM, Baraitser M (1991) Dominantly inherited syndrome comprising partially absent eye muscles, hydrocephaly, skeletal abnormalities and a distinctive facial phenotype. Am J Med Genet 40: 417–420

De Hauwere RC, Leroy JG, Adriaenssens K (1973) Iris dysplasia, orbital hypertelorism and psychomotor retardation: a dominantly inherited developmental syndrome. J Pediat 82: 679–681

Derby GS (1930) Obituary – Theodor Axenfeld, MD. 1867–1930. Arch Ophthalmol 4: 729–732

Obituary – Theodor Axenfeld (1930) Br J Ophthalmol 14: 537–539

Obituary – Prof. Axenfeld (1930. Lancet 2: 429

Rieger H (1935) Beitraege zur Kenntnis seltener Missbildungen der Iris: ueber Hypoplasie des Irisvorderblattes mit Verlagerung und Entrundung der Pupille. Graefes Arch Ophthalmol 133: 602–635

Rieger H (1941) Erbfragen in der Augenheilkunde. Graefes Arch Ophthalmol 143: 277–299

# BAMATTER, Frédéric
*(1899–1988)*

*Courtesy:* Professor D. Klein, Switzerland

B AMATTER syndrome or gerodermia osteodysplastica is a dwarfing skeletal dysplasia characterised by a prematurely aged facial appearance, frequent fractures and spinal malalignment. Inheritance is autosomal recessive.

# Biography

B AMATTER was an eminent academic paediatrician in Geneva during the middle years of the twentieth century.

Frédéric Bamatter was born in Basle, Switzerland on 4 May 1899. He studied at the Universities of Basle, Geneva and Paris, qualifying in medicine in 1927. Bamatter then undertook postgaduate training at the Geneva Institute of Pathological Anatomy, the Lausanne University Hospital and the Pasteur Institute, Paris and he obtained a doctorate in 1932.

Bamatter's wide medical interests became focused on paediatrics and he gained experience in that discipline with Guido Fanconi at Zurich. During this period he identified the poliomyelitis virus as an aetiological agent in bulbar palsy and discovered that meningococcal septicaemia could be implicated in the pathogenesis of failure of the adrenal glands.

In 1935, Bamatter commenced paediatric practice in Geneva and in 1937 he entered into a long-standing collaboration with Professors Adolph Franceschetti and David Klein (*see The Man Behind the Syndrome* p. 58, p. 94), at the Geneva University Ophthalmological clinic. This initiative facilitated the delineation of numerous genetic disorders, including the condition which carries Bamatter's name. In 1951 Bamatter took up a post in Paediatrics at the Geneva University Hospital and thereafter he ascended through the academic ranks, becoming full Professor in 1957. He remained in this post until retirement in 1969 at the age of 70 years, when he was granted honorary professorial status.

Bamatter was one of the founders of neonatology and in 1952 he was instrumental in establishing this speciality at the Geneva maternity hospital. He was subsequently involved in the design of a new Paediatrics Department at the Geneva University Hospital, and he also set up units for childhood neuropsychiatric disorders and for cerebral palsy. Bamatter made many important contributions to paediatrics, including the elucidation of the effects of intrauterine toxoplasmosis infection on the developing foetus. He was responsible for coining the term "embryopathy" in order to distinguish foetal damage due to intrauterine infection from genetically determined malformations. In addition to his work on congenital malformations, Bamatter took special interest in cerebral palsy and in his long career, he published more than 150 medical articles, many of which were concerned with these problems.

Bamatter served as a consultant to the World Health Organization maternal and child welfare section and as chairman of the Swiss Society of Paediatrics. His international status was recognised by awards from universities in France, Italy and Algeria. A special symposium was held in his honour at the Geneva University Paediatric Department on the occasion of his retirement in 1969 and 10 years later his 80th birthday was celebrated in the same manner.

Bamatter died peacefully in Geneva on 25 May 1988 at the age of 90 years.

# Nomenclature

T OGETHER with his colleagues Franceschetti and Klein, Bamatter played a major role in the development of ophthalmological genetics. This highly productive group met regularly in Geneva and amongst the conditions which they documented was gerodermia osteodysplastica, which also bears Bamatter's eponym. They studied 5 affected members of a Swiss family and reported the disorder in the local medical press in 1949 and at an international level in 1950. The similarity of the phenotypic manifestations to those of progeria was emphasised and they commented that their patients resembled the cinema dwarfs created by the film maker, Walt Disney. The term "Walt Disney Dwarfism" enjoyed popularity for some time, but it has now fallen into disfavour.

A further account of the affected Swiss family was published by Klein, Bamatter and colleagues in 1968, and in the following year Boreux, who had been a co-author of the preceding paper, suggested that inheritance might be X-linked recessive, with minor manifestations in hemizygous females. Further reports followed, including accounts of affected children being born after an incestuous relationship and siblings with the condition in an endogamous Mennonite religious isolate. It is now evident that the syndrome is inherited as an autosomal recessive trait.

There has been some debate concerning the terminology and Wiedemann (1978) has argued that as "derma" is neuter the condition should be designated "geroderma osteodysplasticum" or "gerodermia osteodysplastica". There is also some uncertainty about syndromic boundaries because of phenotypic overlap with the De Barsy syndrome (*see* p. 209) and Cutis Laxa with Bone Dystrophy. This issue of homogeneity versus heterogeneity remains unsettled.

# References


Bamatter F, Franceschetti A, Klein D, Sierro A (1949) Gérodermie ostéodysplastique héréditaire. Un nouveau biotype de la "progeria." Confin Neur, Basel 9: 397

Bamatter F, Franceschetti A, Klein D, Sierro A (1950) Gérodermie ostéodysplastique héréditaire. Ann Pédiat 174: 126–127

Boreux G (1969) La gérodermie ostéodysplastie héréditaire (20 ans d'observation). Rev. Oto-Neurogénétique. J Hum Genet 17: 137–138

Klein D, Bamatter F, Franceschetti A, Boreux G, Brocher JEW, Holenstein P (1968) Une affection liée au sexe: la gérodermie ostéodysplastique héréditaire. Rev Oto-Neuro-Ophthalmol 40: 415–421

Klein D (1988) In honor of Prof. Fred Bamatter (1899–1988). J Génét Hum 36(4): 389–391

Oehme J (1992) Fred Bamatter (1899–1988). Kinderkrankenschwester 11(1): 19

Wiedemann H-R (1978) Gerodermia osteodysplastica – What would Virchow have thought about it? Hum Genet 43: 245

# BECKER, Peter Emil

*(b. 1908)*

BECKER muscular dystrophy is characterised by enlargement and weakness of the muscles of the calves. This X-linked condition is comparatively benign with onset in late childhood and slow progression.

# Biography

BECKER is a distinguished German physician and neurologist and was Professor of Human Genetics at Göttingen until his retirement in 1975.

Peter Emil Becker was born in Hamburg on 23 November 1908. He qualified in medicine in 1933 and after training in neurology and psychiatry in Hamburg and Freiburg, became assistant at the Kaiser-Wilhelm Institute of Anthropology, Human Genetics and Eugenics in Berlin.

In 1942 Becker was drafted into the German airforce as a physician and after the war he became a neurologist in Tuttlingen. In 1957 he accepted the chair of Human Genetics at the University of Göttingen and occupied this post for the next 18 years. During this period he edited the *Handbook for Human Genetics* and was a co-editor of the journal *Human Genetics*.

In the early years of his career Becker was involved in population studies and the syndromic delineation of various forms of muscular dystrophy. In the post-war period he discovered the X-linked type of muscular dystrophy which now bears his name. Thereafter, Becker investigated a large series of persons with non-dystrophic myotonia and described several new entities. These include paramyotonia congenita, which derives from a localised area in Westphalia and which formed the subject of a monograph in 1970. An account of Becker's investigations was provided in his autobiographical *Living History*, which was published in 1985. His friend and colleague, Professor Friedrich Vogel of Heidelberg, provided further details of Becker's life and career in his introduction to this article.

In 1937 Becker married Rosett Wendal and the couple had 5 children. Becker retired, with emeritus status, in 1975. He maintained links with his University and in 1993 he was writing a book about the scientists and ideologists who were the founders and precursors of some of Hitler's concepts and beliefs.

# Nomenclature

IN 1955, while practising as a neurologist in Tuttlingen, Becker investigated a German family with an X-linked form of muscular dystrophy. This condition differed from the classical Duchenne dystrophy by virtue of late onset and a benign course. Becker examined 5 affected family members and elicited the information that 6 persons in the kindred had reproduced. The fact that persons with Duchenne dystrophy never procreate was further evidence for syndromic identity. Becker published a brief account of his observations in the German literature, under the title *A new X-chromosomal muscular dystrophy* and his description was further expanded in 1957.

In 1962 Becker wrote a detailed account of a second family in which 10 males were affected. In this article he presented data concerning another kindred which he had found in the archives of the Max Planck Institute, Munich. Becker also mentioned that he had reviewed the literature and recognised the condition in previous descriptions of a family in the Bonn area and another living near Wesel.

Benign X-linked muscular dystrophy has been encountered in many parts of the world and Becker's name has become firmly attached to the condition. It is now evident that the disorder constitutes about 10% of all X-linked muscular dystrophy. Molecular techniques have demonstrated that the genes for Duchenne and Becker dystrophy are allelic and considerable progress has been made in the elucidation of the pathogenesis of the disorder.

# References

Becker PE, Kiener F (1955) Eine neue X-chromosomale Muskeldystrophie. Arch Psychiat Nervenk 193: 427–448

Becker PE (1957) Neue Ergbnisse der Genetik der Muskeldystrophien. Acta Genet Statis Med 7: 303

Becker PE (1985) Living History Biography. Am J Med Genet 20: 699–709

Vogel F (1985) Introduction to PE Becker's Living History · Biography. Am J Med Genet 20: 695–697

# BELL, Sir Charles

*(1774–1842)*

*From:* Duke-Elder S Sir (1971) System of Ophthalmology XII
*Courtesy:* The Wellcome Institute, London

B ELL'S PALSY of the seventh cranial nerve causes asymmetry of the face due to weakness of the facial muscles. The condition may be unilateral or bilateral; closure of the eye is defective, the angle of the mouth drops and there is difficulty in speaking and drinking. There are many causes of facial nerve palsy, including a genetic form which is transmitted as an autosomal dominant trait.

# Biography

S IR CHARLES BELL was a Scottish anatomist, physiologist, neurologist and surgeon who enjoyed a distinguished career in London and Edinburgh during the first half of the nineteenth century. He was a prolific medical writer, a brilliant researcher and a skilled artist.

Charles Bell was born in Doune, Perthshire in 1774. He was a member of a highly talented family; his father and both grandfathers were clergymen, his elder brother John was a surgeon and his 2 other brothers were lawyers. His father died when he was 5 years of age and he received his early education from his mother. At the age of 14 years Bell entered Edinburgh High School and thereafter proceeded to the University of Edinburgh, where he qualified in medicine in 1799. While still a medical student, Bell published a work entitled *System of Dissections* which was illustrated with his own drawings.

Bell became a Fellow of the Royal College of Surgeons of Edinburgh and at the age of 25 years he was appointed to the Edinburgh Royal Infirmary where he continued with his anatomical drawings and the production of wax pathological specimens. Despite his obvious intellectual brilliance, his career was marred by disputes with the Medical Faculty and he moved to London where he taught anatomy in his rooms at Leicester Square. Here he wrote a textbook for artists entitled *On the Anatomy and Physiology of Expression*, which was published by Longmans in 1806. The quality of his work, which attracted royal approval, resulted in his elevation to the Chair of Anatomy at the Royal Academy. An appointment to the staff of the Middlesex Hospital followed and he was awarded a medal for scientific endeavours in the field of neuroanatomy. His surgical interests took him to Brussels in 1815, where he operated on casualties from the Battle of Waterloo. In 1830 Bell published the first modern textbook on neurology, *Nervous Systems of the Human Body*. By this time his reputation as a leading physician was firmly established and in 1831 he was knighted by King William IV.

Bell was a kindly warm hearted man with a quiet sense of humour. He is said to have been as fastidious in his dress as he was meticulous in his dissections. Bell returned to Edinburgh in 1835 to take up an appointment as Professor of Surgery. This period of his life brought him contentment and satisfaction but his happiness at his return to his home city was short lived, as his health failed due to angina pectoris. Bell died in 1842 at the age of 68 years during a visit to the village of Hallow, Worcestershire. Many would agree with his comment "London is a good place to work in, but not to die in."

# Nomenclature

D URING his anatomical studies in London, Bell investigated the pathways and connections of the cranial nerves. His contributions concerning the VIIth cranial nerve were presented to the Royal Society in 1821 and published in the same year. Thereafter, the eponymic form "Bell's palsy" came into general use for facial paralysis due to interference in the lower motor neuron pathway. Upper motor neuron lesions, which affect only the lower portion of the face, do not constitute Bell's palsy.

Familial recurrent facial palsy is well documented and the term "Bell's palsy" was employed by Danforth (1964) in a description of 29 affected persons in a kindred. The eponym was also used by De Santo and Schubert (1969) in a report of ten family members with facial palsy. It has been estimated that about 4% of all cases of Bell's palsy are familial (Yanagihara et al. 1989) and autosomal dominant inheritance with variable expression and low penetrance is well established. A putative autosomal recessive form of episodic Bell's palsy with external ophthalmoplegia was described in four siblings of a consanguineous Jewish kindred (Currie 1970). The family also had diabetes mellitus and as the frequency of Bell's palsy is increased in diabetes the autonomous genetic status of this neurological disorder is unproven. For the sake of semantic accuracy, it must be emphasised that in modern terminology, the term Bell's palsy is reserved for the common idiopathic form of facial paralysis and conditions of known aetiology are specially excluded.

# References

Bell C (1821) On the nerves, giving an account of some experiments on their structure and function which lead to a new arrangment of the system. Phil Trans Roy Soc Lond 111: 398

Corson ER (1910) Sir Charles Bell: The man and his work. Bull Johns Hopkins Hosp 21: 171–182

Currie S (1970) Familial oculomotor palsy with Bell's palsy. Brain 93: 193–198

Danforth HB (1964) Familial Bell's palsy. Ann Otol Rhinol Laryng 73: 179–183

De Santo LW, Schubert HA (1969) Bell's palsy. Arch Otolaryng 89: 700–702

Gordon-Taylor G, Walls EW (1958) *Sir Charles Bell – His Life and Times.* Livingstone, London

Yanagihara N, Yumoto E, Shibahara T (1989) Familial Bell's palsy: analysis of 25 families. Ann Otol Rhinol Laryng 97: 8–10

# BEST, Friedrich

## *(1871–1965)*

*Courtesy:* Professor Dr W. Best, Germany
Professor D. Klein, Switzerland

B EST disease is an autosomal dominant, slowly progressive, degenerative disorder of the macula of the eye. The affected region has a characteristic "scrambled egg yolk" appearance.

# Biography

B EST was a German ophthalmologist who practised in Dresden and Marsberg for more than 50 years.

Friedrich Best was born in 1871 in Wermelskirchen, Germany. He received his ophthalmological training under Leber, and with Bruckner in Darmstadt and Vossius in Giessen. Best settled in Dresden in 1906 and entered ophthalmological practice, heading the Department of Ophthalmology at the Friedrichsstätter hospital. His career was interrupted during World War II, when he lost everything in a massive air raid on Dresden. He retained his energy and drive, and in 1945, at the age of 74 years, set up a new ophthalmological practice in Brilon and Marsberg. Best finally retired in 1960 at the age of 90 years.

Best had wide academic interests, including pathological anatomy, neuropathology, genetics and the physiology of anaesthesia and he published many articles in these fields. In 1901 he discovered a new carmine stain which differentiated retinal rods and cones. This stain was subsequently named "Best's carmine". While working in Giessen in 1905 he delineated the form of macular degeneration which still bears his name. During this period he produced an extensive compilation of the ocular changes in non-inflammatory central nervous system diseases; this was published in the *Kurze Handbuch* and represented the definitive contemporary review of this field. In addition to his abilities as a medical author Best had considerable technical skills and he constructed several pieces of apparatus for anaesthetic and ophthalmological procedures.

Best was a thorough, diligent man, with the virtues of honesty, intellectual integrity and reliability. He was a friend and colleague of many of the earlier generation of German ophthalmologists and especially valued his links amongst the former trainees of the great Leber. He was an honorary member of the Rhine-Westphalian Association of Ophthalmologists and regularly attended the annual congresses in Heidelberg until advanced old age.

Best died in Bonn on 6 June 1965, at the age of 95 years, after a short illness.

# Nomenclature

T HE genetic degenerative disorders of the macula are probably heterogeneous, although syndromic boundaries are ill-defined. The various forms have descriptive and anatomical titles but there is considerable overlap and some semantic confusion.

The polymorphic vitelline form of macular degeneration, to which Best's name is applied, is a well recognised although not necessarily homogeneous entity. Best published his account of this condition in 1905, in a paper on hereditary macular disorders, in which he documented 8 affected family members. The eponym was established by Best's former teacher, Vossius, in a further article on the same family written in 1921. The kindred was studied yet again by Weisel in 1922, and by Jung (1936), when 14 persons in 3 generations were found to be affected. Thereafter the term "Best's disease" was used loosely for any familial macular degeneration.

With the recognition of phenotypic heterogeneity, Best's name was reserved for the form of the disorder known as "vitelline macular degeneration". In 1980, Nordström and Thorburn traced 250 persons in Scandinavia with this condition to a common seventeenth century ancestor. Syndromic identity of Best's disease was confirmed in 1992 when the locus for the faulty gene in this Swedish family was localised to chromosome 11 by a team of ophthalmologists and geneticists at the University of Uppsala.

# References


Best F (1905) Ueber eine hereditaere Maculaaffektion. Z Augenheilk 13: 199–212

Forsman K, Graff C, Nordström S, Johansson K, Westermark E, Lundgren E, Gustavson K-H, Wadelius C, Holmgren G (1992) The gene for Best's macular dystrophy is located at 11q13 in a Swedish family. Clin Genet 42: 156–159

Jung EE (1936) Ber. dtsch. ophthal. Ges., 51, 81 Ueber eine Sippe mit angeborener Maculadegeneration. (Diss.), Giessen (1937)

Nordström S, Thorburn W (1980) Dominantly inherited macular degeneration (Best's disease) in a homozygous father with 11 children. Clin Genet 18: 211–216

Sachsenweger R (1965) Professor Dr. Friedrich Best zum Gedächtnis. Klin Monatsbl Augenheilk 147: 124–125

Vossius A (1921) Ueber die Betsche familiaere Maculadegeneration. Arch Ophthalmol 105: 1050–1059

Weisel (1922) Beitr. z. Bestschen hereditären Maculaerkrankung (Diss.), Giessen

# BIELSCHOWSKY, Max

*(1869–1940)*

*From*: Weil A. (1953) In: Webb Haymaker (ed) The Founders of
Neurology, 1st edn

*Courtesy*: Charles C. Thomas, Springfield, Illinois
Dr Robert Wartenberg, San Francisco, California

**J** ANSKY-BIELSCHOWSKY disease, or late infantile familial amaurotic idiocy, is a progressive neurodegenerative disorder, which leads to death by the end of the first decade. The condition differs from the other forms of amaurotic idiocy by virtue of the lack of fundal changes and the presence of significant cerebellar dysfunction.

# Biography

**B** IELSCHOWSKY was a neuropathologist of the German school. He made notable contributions to the understanding of the pathogenesis of disorders of the central nervous system.

Max Bielschowsky was born in 1869 in Breslau (present day Poland), where his father was a merchant. He studied medicine in his home city and in Berlin and Munich, qualifying with honours in 1893. Bielschowsky had been interested in the brain since childhood, and he trained in neurohistopathology at the Senckenberg Institute in Frankfurt. He was involved in the development of myelin stains and, together with the neurologist Paul Schuster, successfully documented the histological changes in disseminated sclerosis. He also produced a monograph on myelitis and optic neuritis during this period.

In 1904 Bielschowsky was appointed to the Neurobiological Institute of the University of Berlin where he spent the next 30 years. He made significant contributions to the understanding of the normal ultrastructure of the brain and on conditions such as syringomyelia, neurofibromatosis and myotonia congenita. During his productive career he published more than 180 medical articles and he received recognition by election to the fellowship of the Kaiser Wilhelm Gesellschaft. In his personal character Bielschowsky was a quiet, intelligent, diligent man. Although primarily a laboratory scientist, he maintained his clinical interests and was regarded as a competent neurologist.

Bielschowsky – who is sometimes confused with his cousin Alfred Bielschowsky (1871–1940), an ophthalmologist – had an uneasy relationship with Oskar Vogt[1], the director of the Institute, and in 1933 it became necessary for him to resign. In the following years he worked in Utrecht and Madrid, before moving back to Berlin in 1936. An infarction of the cerebellum impaired his health and his life was further complicated by social pressures.

Bielschowsky moved to London with one of his 3 sons in 1939, and in the following year died from a stroke. His ashes are buried in Golders Green Cemetery, next to those of his old friend, Paul Schuster, with whom he had worked in Frankfurt more than 40 years previously.

# Nomenclature

**T** HE group of neurodegenerative disorders, or amaurotic familial idiocies, of which Jansky-Bielschowsky disease is a member, have been the subject of immense nosological confusion.

The history of these disorders, in current understanding, goes back to 1887, when Bernard Sachs of New York described "a familial form of idiocy, generally fatal and frequently associated with blindness". The fundal changes had previously been documented in 1881 by Warren Tay of London and by the turn of the century "Tay-Sachs disease" and "amaurotic familial idiocy" were regarded as being synonymous.

In 1897 Rayner Batten, a London ophthalmologist, documented 2 brothers with macular dystrophy, which commenced at puberty, and in 1903, his brother, Frederick Batten, a London neurologist, described siblings with ocular abnormalities, plus progressive cerebral degeneration (*see The Man Behind the Syndrome* p. 13). Frederick Batten reviewed the disorder in 1910 and in 1915 he discussed the histopathological changes in the eyes and central nervous system, in conjunction with M. S. Mayou. At this stage the eponyms of Tay, Sachs, Batten and Mayou were interchangeably applied to amaurotic familial idiocy.

Bielschowsky contributed to the situation in 1914, when he reported 3 siblings with a progressive disorder which led to dementia and blindness by the age of 10 years. He documented marked cerebellar atrophy and deposition of lipopigment in the neurons. Bielschowsky regarded this condition as a late onset form of Tay-Sachs disease and maintained this concept when he published further findings in 1921.

In the early years of the present century Jan Jansky (1877–1921), a Prague psychiatrist, had reported a family with a similar condition, and the conjoined eponym Jansky-Bielschowsky disease came into use. The nosological situation still remained confused, however, and the eponyms of Bernheimer, Dollinger, Kufs, Stock, Vogt and Spielmeyer were also applied to this group of disorders. In 1958 David Cogan gave a detailed discussion on the eponymous status of the various forms of amaurotic familial idiocy, basing his classification upon the changes in the fundus of the eye. The nosological situation was thereby clarified. Further steps towards resolution have been made by delineation on a biochemical basis.

The Tay-Sachs group of disorders has been separated out and the amaurotic familial idiocies have been termed "neuronal ceroid-lipofuscinoses" with numerical designations to indicate sub-types based upon their age of onset and natural history. Recent molecular studies are strongly suggestive of further residual heterogeneity.

# References


Batten FE (1903) Cerebral degeneration with symmetrical changes in the maculae in two members of a family. Trans Ophthalmol Soc UK 23: 386–390

Batten RD (1897) Two brothers with symmetrical disease of the macula, commencing at the age of fourteen. Trans Ophthalmol Soc UK 17: 48–53

Bielschowsky M (1914) Uber spätinfantile familiäre amaurotische Idiotie mit Kleinhirnsymptomen. Dtsch Zeit Nervenheilk 50: 7–29

Bielschowsky M (1921) Zur Histopathologie und Pathogenese der amaurotischen Idiotie mit besonderer Berücksichtigung der zerebellaren Veränderungen. J Psych Neur 26: 123–199

Cogan DG (1958) Amaurotic family idiocy. A case for eponyms. New Engl J Med 258: 1212-1213

Lewy FH (1941) Max Bielschowsky. Trans Am Neur Assoc 67: 243–244

Ostertag B (1959) In Memoriam, Max Bielschowsky. Dtsch Med Wschr 84: 765–766


---

[1]Oskar Vogt is sometimes mistaken for other persons of the same surname who were active in ophthalmology or neurology.

# BIEMOND, Arie

*(1902–1973)*

*Courtesy*: Dr L. Trotsenburg, Amsterdam

B IEMOND syndrome, or posterior column ataxia, is an autosomal dominant neurological disorder in which disturbance of gait develops in early adulthood.

## Biography

B IEMOND was the doyen of Dutch neurologists during the middle years of the twentieth century.

Arie Biemond was born in Amsterdam on 5 May 1902 and received his schooling and medical education in that city. He developed an interest in neurology while still a student and was eventually appointed as intern to Professor B. Brouwer, the head of neurology at the Wilhelmina Hospital. In 1926 Biemond was promoted to the grade of assistant physician, and in 1929 he obtained his doctorate by successfully defending his thesis on the subject of the cortical connections of the optical system in rabbits and monkeys. Biemond spent his whole career in Amsterdam, succeeding Brouwer in the chair of neurology in 1947 and eventually retiring in 1970.

Biemond was a meticulous and diligent clinician, noted for his thorough and methodical clinical examinations and for his detailed case notes, which were written in his elegant longhand script. In addition to his vast clinical experience, Biemond kept an extensive card index of the neurological literature and in this way he had rapid access to a massive body of information. At the height of his long career, he was widely regarded as the foremost neurologist in Holland and the leader of his speciality in that country.

Biemond was active in neurological research and he was the first to recognise the role of disturbance of potassium metabolism in familial periodic paralysis. He also made contributions to the understanding of ataxia telangiectasia, positional nystagmus, the Guillain-Barré syndrome and the familial disorder of the posterior columns, which now bears his name. He was author or co-author of more than 150 publications and he also wrote textbooks on diseases of the brain and the peripheral nerves; both became the standard works in the Dutch language and the former was also translated into English.

In addition to his clinical skills, Biemond was a fine lecturer. His biographer, Professor Dr J. Bethlem, stated "Nobody in the medical faculty in Amsterdam could teach as he did. Without apparent effort, he could describe a history or demonstrate symptoms in a most vivid and moving way. Patients, demonstration boards, and slides were never lacking. But the effortlessness was deceptive. The lectures were in fact prepared to the smallest detail, often many days in advance and with the assistance of numerous staff members. His innate talents as a speaker and teacher made every lecture an event. He enjoyed a large audience and crowded lecture halls and loved the appreciative remarks heard on the way back to his room when his thoughts were already on the next lecture. His unforgettable lectures and his indestructible enthusiasm prompted many a student to begin specialising in neurology."

Biemond had considerable energy and a prodigious capacity for hard work. He also had profound insight into human nature and he was respected and liked by his staff. After his elevation to professorial status in 1947, Biemond was able to train a group of young neurologists who subsequently headed the specialist divisions of his clinic, which developed an international reputation for excellence.

Despite his high level of activity and his extensive professional contacts, Biemond had a quiet personal lifestyle. His first wife died after a long illness and he himself suffered from a retinal detachment, from which he fortunately fully recovered.

In 1970 when he had been associated with the University of Amsterdam for almost 50 years, Biemond retired because of ill health. He died in Amsterdam on 30 August 1973, at the age of 71 years.

## Nomenclature

B IEMOND'S name has been attached to several genetic neurological disorders and, in the absence of additional descriptive information in their titles, confusion is possible. The addition of numerical designations has failed to resolve this difficulty, as these have been used promiscuously in the literature.

The eponym is most frequently used for a form of posterior column ataxia which Biemond documented in the French literature in 1951. He described a father, brother and 4 offspring with disturbance of posterior column function, which developed in young adulthood and led to disturbance of gait. Biemond's eponym was used in the title of an article published in 1973 in which Singh and colleagues reported a second affected family. This condition is listed in McKusick's Catalogue as "Posterior Column Ataxia", without the asterisk of secure syndromic identity.

The title "Biemond syndrome II" is applied to the combination of iris coloboma, mental retardation, obesity, hypogenitalism and postaxial polydactyly. The syndromic status of this rare condition, which resembles the Laurence-Moon-Biedl-Bardet syndrome, is uncertain.

Biemond congenital and familial analgesia, which comprises diminished pain, touch and temperature sensation, was documented in a male and female twin pair in 1955. This condition, which is rare, may be confused with dysautonomia.

The name of Biemond is sometimes attached to an autosomal dominant syndrome comprising brachydactyly, nystagmus and cerebellar ataxia, which he described in 1934. The eponym is also applied to an autosomal dominant benign distal myopathy, reported by Biemond in 1955. Both conditions are rare and probably represent private syndromes.

## References


Bethlem J (1974) To the Memory of Professor A Biemond. Z Neurol 206: 285–286

Biemond A (1934) Het syndroom van Laurence-Biedl en een aanverwant, nieuw syndrome. Ned Tijdschr Geneesk 78: 1801–1814

Biemond A (1934) Brachydactylie, nystagmus en cerebellaire ataxie als familiar syndrome. Ned Tijdschr Geneesk 78: 1423–1431

Biemond A (1951) Les dégénérations spino-cérébelleuses. Folia Psychiat Neerl 54: 216–223

Biemond A (1955) Investigation of the brain in a case of congenital and familial analgesia. Proc 11th Internat Cong Neuropath, London

Biemond A (1955) Myopathia distalis juvenilis hereditaria. Acta Psychiat Neur Scand 30: 25–38

Singh N, Mehta M, Roy S (1973) Familial posterior column ataxia (Biemond's) with scoliosis. Europ Neurol 10: 160–167

Van Trotsenburg L (1973) Personalia. In memoriam Prof. Dr. A. Biemond. Ned Tijdschr Geneesk 117: 1439–1440

# BLOCH, Bruno
## *(1878–1933)*

*From*: Shelley W.B., Crissey J.T. (1953) Classics in Clinical Dermatology

*Courtesy*: Charles C. Thomas, Springfield, Illinois
Dr Guido Miescher

B LOCH-SULZBERGER syndrome or incontinentia pigmenti is characterised by a disturbance of skin pigmentation associated with lesions of the eyes, teeth, nails and skeleton. Pedigree data are suggestive of X-linked dominant inheritance with lethality in the male (*see* Sulzberger, p. 163).

# Biography

B LOCH was an academic dermatologist in Switzerland during the early decades of the twentieth century.

Bruno Bloch was born on 19 January, 1878, in Endingen, Switzerland. He qualified in medicine at the University of Basel in 1900 and, after training in internal medicine, obtained special experience in dermatology in Berlin, Vienna, Paris and Berne. In 1908 Bloch was appointed as head of dermatology at the University of Basel and in 1916 he was called to the newly established chair of dermatology at the University of Zurich, where he remained until his death in 1933.

Influenced by his mentor, Professor Joseph Jadassohn of Berne, Bloch applied laboratory techniques to the elucidation of dermatological disorders. He made notable contributions in the field of allergy and its role in ringworm and eczema and he had special expertise in the pathogenesis of disorders of pigmentation. Bloch is remembered for his fundamental work on the Dopa reaction for the staining of melanin-forming cells; his junior colleague, Sulzberger, with whom he subsequently shared the eponymous title of the disorder which bears their name, stated in a retrospective article: "Shortly before I received my MD degree from the University of Zurich in 1926, I started working in the Dermatologic University Clinic of Professor Bruno Bloch. While Bruno Bloch was at the University of Basel, he discovered and developed the Dopa reaction to stain, specifically those mammalian cells that had the capacity to form melanin pigment. This discovery was not merely a new histologic staining method – it was an entirely novel, fundamentally different technique. Bloch's discovery made possible the demonstration of a dynamic process or specific capacity within the cells and pin-pointed the intracellular organelles which were endowed with particular capacity – in this case, the capacity of melanogenesis."

Bloch was an enthusiastic teacher with an enquiring scientific mind and he provided inspiration for many of the students who attended his clinics.

Bloch's career ended prematurely in 1933 when he died at the early age of 55 years.

# Nomenclature

T HE delineation of incontinentia pigmenti, which is now also know as the Bloch-Sulzberger syndrome, can be attributed to Garrod (1906) who wrote an account of a mentally retarded girl with tetraplegia and the characteristic skin pigmentation. In 1925 Bardach published a similar report in the German literature, as did Bloch in 1926. Bloch stated: "A two year old girl; the affection has been present since birth and has apparently not changed much since that time. The pigmentary changes are most pronounced on the lateral parts of the trunk, nearly symmetrical toward the midline anteriorly and posteriorly, forming irregular rings about the mammae, and from the right axilla out to the middle of the forearm, as well as irregular bands from the hips to the ankles. The colour of the affection is very distinctive, not a true brown as in pigmented nevi, but rather a dirty brown with a distinct tinge of slate grey, the margins a dirty light yellowish colour, especially on the back. The form of the pigment spots is most remarkable. They occur in completely irregular splashes and show figures with spidery projections such as have not heretofore been described in any pigment anomaly. The projecting arms of individual spots join with one another in various ways. The whole picture has something capricious and artificial about it, as if someone had painted completely irregular patterns on the skin."

Sulzberger reported a further case in 1928 and in the following year, Siemens described "a new pigmentary dermatosis" which also seemed to represent the same disorder. Bloch had used the term "incontinentia pigmenti" in the title of his paper and this descriptive designation was subsequently accepted with the frequent attachment or alternative use of the conjoined eponym "Bloch-Sulzberger syndrome".

# References


Bardach M (1925) Systematisierte Naevusbildungen bei einem eineiigen Zwillingspaar. Z Kinderheilkd 39: 542–550

Bloch B (1926) Eigentumliche, bisher nicht beschriebene Pigmentaffektion (incontinentia pigmenti). Schweiz Med Wochenschr 7: 404–405

Garrod AE (1906) Peculiar pigmentation of the skin in an infant. Trans Clin Soc Lond 39: 216

Obituary (1933) Br J Dermatol 45: 269

Sulzberger MB (1928) Ueber eine bisher nicht beschriebene congenitale Pigmentanomalie (incontinentia pigmenti). Arch Dermatol Syph (Berl) 154: 19–32

Sulzberger MB (1980) Three lessons learned in Bloch's clinic. Am J Dermatopath 2(4): 321–325

# BLOUNT, Walter Putnam
*(b. 1900)*

B LOUNT disease, tibia vara or osteochondrosis deformans tibiae, presents as bow legs in childhood. The pathogenesis is obscure and heterogeneity is likely; most cases are sporadic but involvement of successive generations has been recorded.

# Biography

B LOUNT was an orthopaedic surgeon in Wisconsin, USA. He played a crucial role in the development of the Milwaukee spinal brace.

Walter Blount was born on 3 July 1900 in Oak Park, Illinois, USA, where his mother was a medical practitioner and his father a teacher and scientific author. He achieved academic and sporting distinction at the University of Illinois before qualifying at the Rush Medical College in 1925. Blount trained in orthopaedic surgery at the Wisconsin General Hospital and after obtaining additional experience overseas he was appointed to the staff of the Milwaukee Children's Hospital. In 1957 he became professor of orthopaedics at the Marquette Medical School and occupied this post until his retirement.

Blount made a significant contribution to the practice of modern paediatric orthopaedics by virtue of his role in the development of the Milwaukee brace for spinal malalignment. In addition he published a classical work on childhood fractures in 1954 and wrote a second edition in 1977. His achievements were rewarded by election to honorary membership of several international orthopaedic associations, including presidency of the American Academy of Orthopaedic Surgeons in 1954 and Vice-Presidency of SICOT in 1966.

During an academic visit to Europe, Blount married Frances Hobden, the daughter of the president of Kalamazoo College, USA. The couple had 2 children, Ralph and Jane, and 5 grandchildren. After the death of his first wife in 1966, Blount married a neighbour, Jane Dunlok Telander. After formal retirement Blount became emeritus professor. He enjoyed walking, cycling and gardening, in addition to lecturing and writing. He also maintained contact with the faculty and residents and continued his academic activities and consulting practice for many years.

# Nomenclature

I N 1922, Erlacher documented progressive deformity of the upper end of the tibia, using the term "tibia vara" in his text. In 1937 Blount reported 13 children with a form of bow legs which he termed "osteochondrosis deformans tibiae". The condition was the consequence of deformation of the upper medial tibial epiphyses and metaphyses but the aetiology was unknown. Further reports followed under either the eponym or the descriptive title and it became apparent that the disorder was heterogeneous, "infantile" and "late" forms being recognised. Blount retained a life-long interest in the disorder, and in 1966, 30 years after his original paper he wrote a review of the condition in the authorative *Current Practice in Orthopaedic Surgery*. The term "Blount disease" is in general use, although there is considerable confusion concerning the different forms of the condition.

Although Blount disease is supposedly rare, the infantile form is common amongst the indigenous population of South Africa. Indeed, in a 10-year period, more than 300 affected children were treated at the Baragwanath Hospital, Johannesburg (Bathfield and Beighton 1978). The majority of occurrences are sporadic but in 1957 Tobin observed the disorder in a father and his 2 sons, while Silbert and Bray (1977) described the condition in 4 generations of a family. This autosomal dominant form of Blount disease seems to be a genuine entity but is rare and makes up only a small proportion of all examples of the disorder.

# References


Bathfield CA, Beighton PH (1978) Blount disease. A review of etiological factors in 110 patients. Clin Orthop 135: 29–33

Blount WP (1937) Tibia vara: osteochondrosis deformans tibiae. J Bone Joint Surg 19A: 1–29

Blount WP (1966) Tibia vara: osteochondrosis deformans tibiae. In: Current Practice in Orthopaedic Surgery Vol 3 pp. 141. CV Mosby Co. St. Louis, Illinois.

Erlacher P (1922) Deformierende Prozesse der Epiphysengegend bei Kindern. Archiv für Orthopädische und Unfall-Chirurgie 20: 81–96

Silbert JR, Bray PT (1977) Probable dominant inheritance in Blount's disease. Clin Genet 11: 394–396

Tobin WJ (1957) Familial osteochondritis dissecans with associated tibia vara. J Bone Joint Surg 38A: 1091–1105

# BONNEVIE, Kristine Elisabeth Heuch

*(1872–1948)*

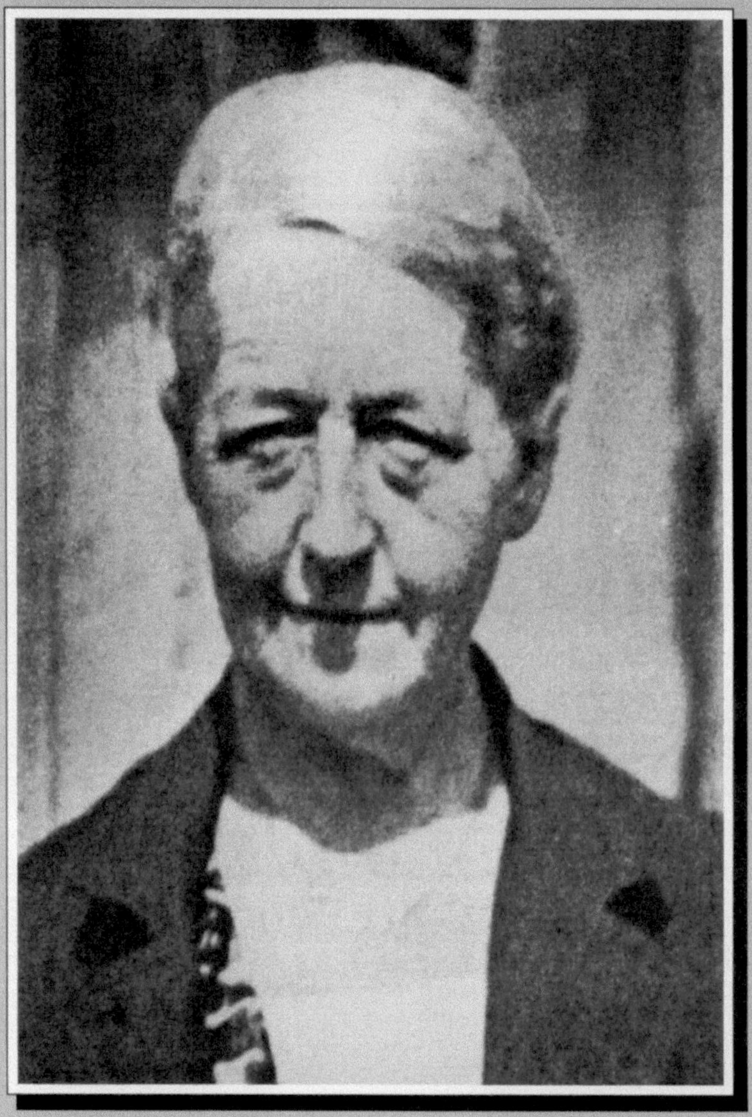

*From*: Yearbook of the Norwegian Academy of Science and
Letters (1949)

*Courtesy*: Mr Atle Vesterjo, The Norwegian Academy of Science
and Letters
Dr Anton Brogger, Norway

# B

ONNEVIE-ULLRICH-TURNER syndrome (*see* Ullrich, p. 169).

## Biography

**B**ONNEVIE was a developmental biologist, cytogeneticist and human geneticist during her long career as professor of zoology at the University of Oslo, Norway.

Kristine Bonnevie was born on 8 October 1872 in Trondheim, where her father, Jacob Aall Bonnevie (1838–1904) was a teacher and cabinet minister. After she had completed her schooling, she entered the University of Oslo, obtaining an appointment in 1900 as curator of the museum of zoology and proceeding to her doctorate in 1906. Bonnevie became the first female professor in Norway when she succeeded to the chair of zoology in 1912.

Bonnevie's early work, in the closing years of the nineteenth century, included the appraisal of ascidiae and hydroidae specimens collected by the Norwegian North Oceans Expedition. After obtaining postgraduate experience with Lang in Zurich and Boveri in Wurzburg, she commenced her doctoral project, which concerned the development of germ cells of a parasitic snail. The investigation led Bonnevie into a life-long interest in chromosomal structure and function. In 1911 Kristine Bonnevie became the first female member of the Norwegian Academy of Science and Letters when she was elected to the Science Class.

Bonnevie became involved with human genetics in 1912, when she instituted a large-scale study of dwarfism, polydactyly and twinning in endogamous, isolated communities in mountainous and fiord regions. By 1916 she had established a University Institute for Research on Heredity, of which she remained director until her retirement. Within this Institute Bonnevie undertook extensive investigations into dermatoglyphics and thereby established an international reputation in human genetics. In 1927 Bonnevie became interested in gene action during embryogenesis, using hereditary defects in mice as her experimental model. She retired from her chair in 1937, at the age of 65 years, but continued with her research in the Zoology Laboratory. During her long and productive life she published many scientific articles, the last being submitted shortly before her death in 1948.

Bonnevie was a popular teacher, with a reputation as an excellent lecturer, both in her own field and in popular science. Despite her seniority, she was much concerned with the welfare of her students and in 1920 she received a gold medal for her services in this respect. Bonnevie had a forceful character and a strong sense of civic duty; she served as a member of the Oslo City Council, a deputy member of Parliament and on the Norwegian delegation to the League of Nations in Geneva. In 1946 she was honoured by appointment as Knight of the Order of St Olav for her courage during the war in connection with the organisation of food supplies for the Resistance and for her students.

Bonnevie died in 1948 at the age of 76 years. In a eulogy given at a meeting of the Norwegian Academy for Science and Letters, Bjorn Foyn quoted her personal philosophy: "Age and death follow as natural parts of the life of each subject – in the same way as the plants wither at the end of their flowering period. The individual has done its deeds, and life is at an end. But if they have succeeded during their lifetime in arriving at some of the goals of the ethics of Nature, to live according to the best in their characters, then their lives will, without doubt, leave some marks behind among their fellows and relatives."

A building at the University of Oslo now carries the name of Kristine Bonnevie.

## Nomenclature

**I**N the decade preceding World War II, Bonnevie was much involved with developmental defects in mice and she used the myelencephalic bleb mouse of Bagg and Little for her studies of abnormal embryogenesis. Lymphoedema of the neck was a major manifestation in these mice and, in the light of Ullrich's earlier work in this field, the term "Status Bonnevie-Ullrich" (*see* Ullrich, p. 169) came into use. Initially, this designation pertained to malformation of the tissues of the neck, and for many years it was the subject of considerable controversy amongst developmental biologists. In an editorial review, Opitz (1991) pointed out that luminaries who were involved in the debate, which sometimes became contentious, included Cotterman, Ruggles, Gates, Engel and Julius Bauer.

With the growing interest in medical genetics and human malformation, the appellation "Bonnevie-Ullrich" was applied to any form of nuchal webbing which followed localised lymphoedema. This designation then evolved into a name for a group of disorders, which had not been clearly delineated at the time, but which comprised females with gonadal dysgenesis with an XO chromosomal constitution (now known as the Turner syndrome or Ullrich-Turner syndrome), together with males with stunted stature, neck webbing and structural cardiac defects (now termed the Noonan syndrome, *see* p. 121). In 1965 Ferguson-Smith clarified this issue to some extent when he defined gonadal dysgenesis in terms of karyotype-phenotype correlations. When the separate syndromic identity of these disorders was recognised, the eponym "Bonnevie-Ullrich" was applied to the "male Turner" syndrome. Noonan drew the attention of medical geneticists to this disorder and her name then replaced the conjoined eponym of Bonnevie and Ullrich. The terminology became even more complicated, as Bonnevie's name was also sometimes applied to an ill-defined disorder of isolated neck webbing, with or without abnormalities in other regions. The independent syndromic status of this "pterygium colli" condition is very doubtful.

The nosological situation has now been clarified and the term "Turner syndrome" or "Ullrich-Turner syndrome" is applied to females with gonadal dysgenesis and an XO chromosomal constitution. Noonan's name is reserved for males with stunted stature, variable neck webbing, structural cardiac defects and normal chromosomes. Sadly, Bonnevie's eponym has largely fallen into disuse.

## References

Bonnevie K (1934) Embryological analysis of gene manifestation in Little and Bagg's abnormal mouse tribe. J Exp Zool 67: 443–520

Ferguson-Smith MA (1965) Karyotype-phenotype correlations in gonadal dysgenesis and their bearing on the pathogenesis of malformations. J Med Genet 2: 142–147

Foyn B (1949) Eulogy. Meeting of Norwegian Academy for Science, 30 September

Opitz JM (1991) Otto Ullrich: an appreciation. Am J Med Genet 41: 126–127

Rossi E, Caflisch A (1951) Le syndrome du pterygium status Bonnevie-Ullrich, dystrophia brevicolli congenita, syndrome de Turner et arthromyodysplasia congenita. Helv Paediatr Acta 6: 119–148

Ullrich O (1930) über typische Kombinationsbilder multipler Abartungen. Zchr Kindh 49: 271–276

# BOURNEVILLE, Désiré Magloire

*(1840–1909)*

*From*: Benda C.E. (1953) In: Webb Haymaker (ed) The Founders
of Neurology, 1st edn

*Courtesy*: Charles C. Thomas, Springfield, Illinois
Portrait from: Nos grands médecins d'aujourd'hui, Paris 1891; by
H. Bianchon

**B**OURNEVILLE disease, or tuberous sclerosis, is a well-recognised autosomal dominant disorder, in which cutaneous and intracranial lesions are variable components. Epilepsy and mental retardation are frequent complications.

# Biography

**B**OURNEVILLE was a nineteenth century French physician. He made notable contributions to paediatric psychiatry and pioneered the concept of special day schools for mentally handicapped children.

Désiré Bourneville was born in 1840 in the village of Garancières, Eure, Normandy, where his father was a minor landholder. He qualified in medicine in Paris and shortly afterwards distinguished himself during a cholera epidemic in Amiens.

Bourneville served as a surgeon in the 160th Battalion of the National Guard during the Franco-Prussian war and subsequently as assistant medical officer at a field hospital in the Jardin des Plantes, during the siege of Paris. In 1871, during the Commune, Bourneville showed great courage in preventing revolutionaries from murdering their wounded enemies who were in his care.

In 1872 Bourneville became editor of Charcot's *Leçons sur les maladies du système nerveux faites à la Salpêtrière* and in the following year, founded the journal *Le Progrès Médicale*. Bourneville was amongst the few persons held in any esteem by the great Charcot and, in 1880, with the latter's support, he launched the important journal *Archives de Neurologie*.

Bourneville was a brilliant scholar and had an international reputation as an expert in childhood mental retardation. His main academic interests were in the field of paediatric neurology and psychiatry and he collaborated with Pascal, Bernard and his mentor, Delasiauve, in these disciplines. He published numerous papers on mental retardation and epilepsy and documented the clinical implications of hypothyroidism and mongolism. He also established tuberous sclerosis as a syndromic entity (vide infra).

Bourneville held an appointment as paediatric physician at the Bicêtre Hospital from 1879 until his retirement in 1905. During this period he founded a day school for handicapped children; this concept was subsequently followed in many other countries throughout the world. Bourneville took a profound interest in his little patients, and it is recorded that "on Saturdays he held open house at the Bicêtre in which his charges performed exercises and dances to the accompaniment of a band composed of idiots, epileptics and spastics; the trombonist had wooden legs".

Bourneville had considerable strength of character and the courage of his convictions. In addition to his medical activities, he held public office in Paris in the 1870s and 1880s; in this role his reformist zeal frequently precipitated political conflict. Nevertheless, his standing was not diminished and he was mourned by the Government and the public when he died in 1909.

# Nomenclature

**I**N 1882 Bourneville reported the pathological changes in the cerebrum of an affected person, using the term "tuberous sclerosis" to describe the potato-like appearance of the abnormalities. Although he considered that these features were the consequence of meningo-encephalitis, he noted the presence of renal tumours and the facial lesions of adenoma sebaceum. It is of interest that he published his article in the first volume of the journal *Archives of Neurology*, which he himself had founded.

Adenoma sebaceum, in which red, warty lesions are characteristically present on the cheeks and nose, was reviewed in detail by Balzer and Ménétrier (1885) and Pringle (1890). These authors regarded this skin disorder as a specific entity, and they failed to appreciate its significance as a syndromic component of tuberous sclerosis.

In a discussion of the pathogenesis of mental retardation, Vogt (1908) employed the designation "tuberous sclerosis" and drew attention to the syndromic association of adenoma sebaceum, mental deficiency and epilepsy. Sherlock (1911), in a work entitled *The Feeble Minded* coined the term "epiloia" to embrace these major features of the disorder. In 1924 Lind used "epiloia" as the title of an article and thereafter this name gained general acceptance. In the same period, the eponym "Bourneville" was employed in the ophthalmological literature by Van der Hoeve (1923). The terms "tuberous sclerosis" and "epiloia" have persisted in the English language literature, although the latter is now falling out of favour. The eponym "Bourneville" was mainly used by Continental authors but it is now rarely encountered.

Current interest in tuberous sclerosis is focused upon chromosomal localisation of the faulty gene and the issues of new mutation, variable clinical expression and possible non-penetrance. Molecular technology has provided new insights into these problems and it is anticipated that there will be significant advances in the understanding of the disorder in the foreseeable future.

# References


Balzer F, Ménétrier P (1885) Étude sur un cas d'adénomes sébaces de la face et du cuir chevelu. Arch Physiol 6: 564

Bourneville DM (1880) Scléreuse tubéreuse des circonvolutions cérébrales. Idiotie et épilepsie hémiphlégique. Arch Neurol, Paris I: 81–91

Lind WAT (1924) Epiloia. Med J Austral 2: 290–294

Noir (1909) Obituary. Progr Méd, Paris 25: 293–295

Pringle JJ (1890) A case of congenital adenoma sebaceum. Br J Derm 1: 1–14

Sherlock EB (1911) *The Feeble Minded*. London, MacMillan.

Thulié (1909) Obituary. Rev Philanthrop, Paris 25: 174–180

Van der Hoeve J (1923) Augengeschwülste bei der tuberösen Hirnsklerose (Bourneville) und verwandren Krankheiten. Arch F Ophthalmol (Berl) cxi: 1–16

Vogt H (1908) Zur Pathologie und pathologischen Anatomie der verschiedenen Idiotieformen. II Tuberose Sklerose. Mschr Psychiat Neurol 24: 106

# BUSCHKE, Abraham

## *(1868–1943)*

*From:* Shelley W.B., Crissey J.T. (1953) Classics in Clinical
Dermatology

*Courtesy:* Charles C. Thomas, Springfield, Illinois
Dr Helen O. Curth

BUSCHKE-OLLENDORFF syndrome or dermatofibrosis lenticularis disseminata (*see* p. 125).

# Biography

**B**USCHKE **was an eminent German dermatologist in Berlin during the first three decades of the twentieth century.**

Abraham Buschke was born in Nakel, Poland in September 1868. He qualified in medicine at the University of Berlin in 1891, after undergraduate studies in Breslau and Greifswald, and subsequently trained in dermatology under Neisser of Breslau and Lesser of Berlin. Buschke spent his professional career in Berlin, becoming instructor in dermatology at the Friedrich Wilhelm University in 1900 and head of dermatology at the Virchow Hospital in 1906. He retired in 1933 at the age of 65 years.

Buschke had drive, energy and a strong personality. In an obituary, his former colleagues, William and Helen Curth (née Ollendorff, see p. 125) stated "his assistants loved his great vitality, although it was not always easy for them to work out to his satisfaction dermatological problems which he handed to them in the morning in the form of cryptic notes scribbled down during the night".

At the height of his career, Buschke had access to 400 beds for patients with dermatological disorders and published numerous articles based upon the vast quantity of clinical material that was available to him. He was involved in the use of thallium in animal experiments and greatly concerned with dermatological aspects of immunological responses to syphilis. At various times in his career, his interests extended to anthropology, chemistry and homeopathy.

As with many of his colleagues, Buschke became victim of the Nazis and together with his wife, he was incarcerated in the Theresienstadt concentration camp, Czechoslovakia. He died in 1943 at the age of 75, but news of his death only reached the outside world two years later, following the release of his widow. Buschke is survived by three sons in the USA, two of whom are medically qualified.

# Nomenclature

**B**USCHKE was a prolific author who wrote definitive descriptions of several entities. His name is attached to scleredema adultorum, disseminated cryptococcosis (Busse-Buschke disease) and giant condyloma acuminatum (Buschke-Lowenstein tumour). In 1928, in conjunction with his junior colleague, Helen Ollendorff, he delineated an unusual disorder of the skin and skeleton, which was termed "dermatofibrosis lenticularis disseminata with osteopathia condensans disseminata". Ollendorff subsequently married Buschke's assistant, William Curth, and in 1934 she published a review of the disorder, using her married name, and documented familial occurrence.

The conjoined eponym "Buschke-Ollendorff" was first employed by Schimpf et al. (1970), in conjunction with the descriptive title of the condition, and thereafter by Verbov (1977), in the same manner. Further reports followed, and the syndromic status of the condition is now well established. Current interest is focused upon the skin lesions as a source of elastin for molecular investigations.

# References


Buschke A, Ollendorff H (1928) Ein Fall von Dermatofibrosis lenticularis disseminata und Osteopathia condensans disseminata. Derm Wschr 86: 257–262

Curth HO (1934) Dermatofibrosis lenticularis disseminata and osteopoikilosis. Arch Derm Syph 30: 552–560

Curth W, Curth HO (1945) Abraham Buschke, M.D. 1868–1943. Arch Derm Syph 52: 32

Obituary (1946) Abraham Buschke, M.D. Br J Dermatol 58: 32

Schimpf A, Roth W, Kopper J (1970) Dermatofibrosis lenticularis disseminata mit osteopoikilie. Buschke-Ollendorff-Syndrome. Dermatologica 141: 409–420

Verbov J (1977) Buschke-Ollendorff syndrome (disseminated dermatofibrosis with osteopoikilosis). Br J Derm 96: 87–90

# CALVÉ, Jacques

*(1875–1954)*

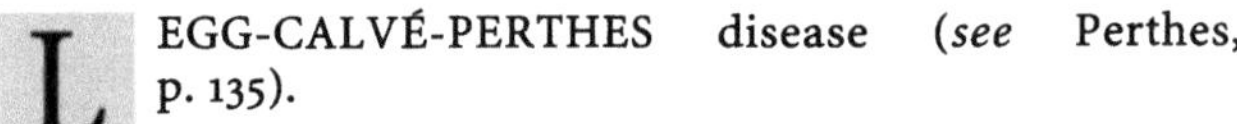

L EGG-CALVÉ-PERTHES disease (*see* Perthes, p. 135).

# Biography

C ALVÉ **was a distinguished French orthopaedic surgeon, noted for his work on tuberculosis of the bones and joints.**

Jacques Calvé was born in France in 1875. He qualified in medicine, trained in orthopaedic surgery in Paris and then took up an appointment at the Maritime Hospital, Berck Plage, on the northern coast of France. This hospital, which had 1100 beds, was dedicated to the long-term care of patients with bone and joint tuberculosis, many of whom came from the Paris area. The surgical director, Ménard, had greatly influenced the orthopaedic management of tuberculosis and after his arrival Calvé was the author of a series of articles in which these principles were expounded.

Calvé succeeded to the surgical directorship of the hospital after the retirement of Ménard and he then set about raising funds for the establishment of a new hospital and a trade school for tuberculous children. The American residents of France made major donations, and the new hospital was duly named "La Foundation Franco-Americaine de Berck". Calvé's interaction with the American community was also reflected in his marriage to the daughter of an officer in the US army.

Calvé was well known in international orthopaedic circles and his description of vertebral osteochondritis in 1925 and his operative approach to intraspinal tuberculous abscesses attracted special attention. He also had a long-standing friendship with the doyen of British orthopaedic surgery, Robert Jones. Calvé's international reputation was enhanced by his dignified manner, personal charm and a kindly disposition.

Following World War II, Calvé lived for some time in the USA. He eventually became unwell and returned to France in 1953, dying in his old hospital at Berck at the age of 79 years.

# Nomenclature

C ALVÉ'S appointment at Berck coincided with the purchase of radiological apparatus and he took the opportunity to obtain radiographs of 500 children who were being treated for tuberculosis of the hip joints. Ten of these children turned out to have non-tuberculous flattening and fragmentation of the femoral capital epiphyses; in his article, published in July 1910, Calvé termed this disorder "pseudocoxalgia". The condition was also documented independently in February 1910 by Legg in the USA and in October 1910 by Perthes in Germany, and the triple eponym came into use. In 1920, Waldenström introduced the term "coxa plana", and the quadruple eponym subsequently enjoyed some favour. Many other descriptive names have been used, including aseptic necrosis of the femoral head, coxalgia infantilis seu infantilis, malum coxae, osteoarthritis coxae, osteochondrosis deformans juvenilis coxae, osteochondropathia deformans juvenilis coxae, osteochondrosis deformans coxae juvenilis, and osteochondrosis of capitular epiphyses of femur. With the move away from descriptive Latin in terminology and for the sake of clarity and expediency, the simple eponymic form "Perthes disease" is now generally employed.

# References


Calvé F (1910) Sur une forme particulière de pseudocoxalgie greffée. Sur une déformation caractéristique de l'extrémité supérieure du femur. Rev Chir (Paris) 42: 54–84

Calvé F (1925) Sur une affection particulière de la colonne vértebrale chez l'enfant simulant le mal de Pott. Ostéo-chondrite vertébrale infantile. J Radiol 9: 22–27

In Memoriam (1954) Jacques Calvé 1875-1954. J Bone Jt Surg 36B: 502–503

Legg AT (1910) On obscure affection of the hip joint. Boston Med SJ 162: 202–204

Perthes GC (1910) Über arthritis deformans juvenilis. Deut Zschr Chir 107: 111–159

Waldenström H (1920) Coxa plana. Osteochondritis deformans coxae, Calvé-Perthe'sche Krankheit, Legg's Disease. Z Chir 47: 539–542

# CANAVAN, Myrtelle M.

*(1879–1955)*

*From:* Arch Neurol Psychiat (1931) 25: 299–308

*Copyright:* 1931 American Medical Association

C ANAVAN-VAN BOGAERT disease is a progressive, degenerative disorder of the central nervous system, which is characterised by "spongy" changes in the white matter. Inheritance is autosomal recessive.

# Biography

C ANAVAN was a North American pathologist in the early decades of the twentieth century. She was one of the first woman to specialise in neuropathology.

Myrtelle May Canavan, née Moore, was born on 24 June 1879 in St Johns, Michigan. She studied at the Michigan State College, the University of Michigan and the Woman's Medical College of Pennsylvania, obtaining her medical qualification in 1905. She then married a medical colleague, James Francis Canavan, and in 1907 obtained a laboratory post at the Danvers State Hospital, Hawthorne, Massachusetts. Here she met Professor Elmer Southard, the head of neuropathology at Harvard medical school and the Boston Psychiatric Hospital, whose influence initiated her life-long interest in neuropathology.

In 1910 Canavan was appointed as resident pathologist at the Boston State Hospital and, in conjunction with Southard, she published a number of articles on clinicopathological correlations in neurological diseases. Canavan became pathologist to the Massachusetts Department of Mental Disease in 1914 and during the next decade, while in this post, she undertook extensive investigations into the neuropathological changes in psychiatric disorders. Numerous publications emanated from this endeavour and she established an international reputation in her field.

In 1924 Canavan became associate professor of neuropathology at Boston University school of medicine and 2 years later she took up an appointment as instructor in neuropathology at the University of Vermont. She also became curator of the Warren Anatomical Museum at Harvard medical school, maintaining links with this museum until her retirement in 1945. Canavan developed Parkinson disease, and she died in Boston on 26 August 1953, at the age of 74 years.

# Nomenclature

I N 1931, while teaching neuropathology at the University of Vermont School of Medicine, Canavan documented the clinical and histopathological features of a child with macrocephaly, marked developmental delay and visual disturbance who died at the age of 16 months. Histological studies revealed spongy degeneration of the white matter, with some preservation of the glia. Canavan described these changes as "lacy oedema" and stated that there were no histopathological pointers to any specific aetiology. A certain amount of nosological confusion was engendered, as Canavan had entitled her article "Schilder's encephalitis periaxialis diffusa"; at that time Schilder's name was also associated with diffuse cortical sclerosis, from which several separate disease entities have now been defined (*see The Man Behind the Syndrome* Schilder, p. 158–159).

More than 15 years after Canavan's article was published, Ludo van Bogaert (*see* p. 171) and Ivan Bertrand reported a number of children with spongy degeneration of the brain. The triple eponym then came into use, although the single appellation "Canavan" is now generally preferred. The nosological situation with regard to Schilder's disease was clarified in 1967, when van Bogaert and Bertrand published a monograph entitled *Spongy Degeneration of the Brain in Infancy*.

Sets of siblings with normal parents were reported and it became evident that Canavan disease was an autosomal recessive disorder, which predominated in children of Ashkenazi Jewish stock (Mahloudji et al. 1970). As case reports accumulated, Canavan disease was categorised into congenital, infantile and late-onset forms. There is considerable clinical overlap and the separate status of these forms of the disorder is questionable, although they all share the nonspecific histological feature of spongy degeneration of the white matter.

In 1988 Matalon and colleagues demonstrated that the basic defect in Canavan disease is deficiency of the enzyme aspartoacylase. The faulty gene has been localised to chromosome 17.

# References


Canavan MM (1931) Schilder's encephalitis periaxialis diffusa: Report of a case in a child aged 16 and one-half months. Arch Neur Psych 25: 299–308

Matalon R, Michals K, Sebesta D, Deanching M, Gashkoff P, Casanova J (1988) Aspartoacylase deficiency and N-acetylaspartic aciduria in patients with Canavan disease. Am J Med Genet 29: 463–471

Mahloudji M, Daneshbod K, Karjoo M (1970) Familial spongy degeneration of the brain. Arch Neurol 22: 294–298

Obituary (1953) JAMA 153(14): 1295

Van Bogaert L, Bertrand I (1967) *Spongy Degeneration of the Brain in Infancy.* Elsevier North-Holland, Amsterdam

# COGAN, David Glendinning

### (1908–1993)

C OGAN congenital ocular motor apraxia is an autosomal recessive disorder in which jerky movements of the head result from impairment of fixation of gaze due to defective cortical control of movements of the external muscles of the eyes.

# Biography

C OGAN was a North American ophthalmologist and an international authority in his field. He was a founder of the subspeciality of neuro-ophthalmology and had a distinguished career at Harvard Medical School and the Massachusetts Eye and Ear Infirmary.

David Glendinning Cogan was born in 1908 in Fall River, Massachusetts, USA, where his father was an Episcopal minister and his mother was a practising ophthalmologist. He attended local schools until 1925, when he enrolled at Dartmouth College, where he combined a collegiate schedule with the first 2 years of medical school. He then transferred to Harvard Medical School, from which he graduated in 1932. After an internship in medicine at the University of Chicago, he undertook a residency in ophthalmology at the Massachusetts Eye and Ear Infirmary and then obtained further experience in Switzerland, Germany and Holland as Mosely travelling fellow. He was appointed to the staff of the Infirmary in 1937 as a clinician with special interests in neuro-ophthalmology and ophthalmic pathology.

Cogan became Director of the Howe Laboratory of Ophthalmic Research, a Senior Surgeon at the Infirmary and the Henry Williams Professor of Ophthalmology at the Harvard Medical School. On his mandatory retirement in 1973 at the age of 65 years, Cogan moved to the National Institutes of Health, where he continued his research activities. Despite his age, the tempo of his work continued unabated and he often put in 10-hour days and 6-day weeks.

Cogan was an accomplished medical author, writing more than 500 articles and several books. In particular, his monograph *Neurology of the Visual System* published in 1966 was a major contribution which became the standard text of its time. His literary skills were also brought into play when he served as chief editor of the *Archives of Ophthalmology* for the period 1960–1966. His research interests included the pathology and physiology of the cornea, and the relationship between central nervous system lesions and oculo-motor disturbances. He was the first to recognise the role of atomic radiation in the causation of cataracts in the survivors of the nuclear explosions at Hiroshima and Nagasaki.

Cogan's academic endeavours were recognised by awards ranging from the Warren Prize in 1944 to the British Mackenzie medal in 1968. In 1985, when he was honoured by the dedication of the David G Cogan Library at the National Eye Institute, his career was outlined in the following way: "So great have been Dr Cogan's achievements in all three areas – research, patient care and education – that he, more than anyone else, is credited with transforming the field of ophthalmology from a branch of surgery into a medical speciality at the forefront of science".

Although Cogan was a modest man, he adhered to that which he believed to be right. Beneath his quiet demeanour he had an irrepressible sense of humour. He was devoted to his family and, outside his work, his main hobby was playing the piano.

Cogan was stricken with a heart attack while returning to Bethesda from his summer home in Leland, Michigan, and he passed away on 9 September 1993 at the age of 85 years.

# Nomenclature

C OGAN delineated the genetic condition which bears his eponym in an article published in 1952. He employed the descriptive term "congenital ocular motor apraxia" and drew attention to the way in which the condition presented with jerky head movements. The autosomal recessive nature of the condition was suggested when Sachs (1967) reported affected siblings with consanguineous parents. Cogan's eponym was used in the title of this article and in that of a review by Vassella et al. which followed in 1972. The condition is rare and by 1994 only about 40 cases had been documented.

In addition to familial ocular motor apraxia, Cogan's name is attached to a syndrome of acute onset nonsyphilitic interstitial keratitis with vestibuloauditory symptoms, which he documented in 1945, and to a form of microcystic corneal dystrophy which he described in 1964. Neither of these disorders seems to have a genetic basis.

# References


Cogan DG (1945) Syndrome of nonsyphilitic interstitial keratitis and vestibuloauditory symptoms. Arch Ophthalmol 33: 144–149

Cogan DG (1952) A type of congenital ocular motor apraxia presenting jerky head movements. Trans Am Acad Ophthal Otolaryng 56: 853–862

Cogan DG, Donaldson DD, Kuwabara T, Marshall D (1964) Microcystic dystrophy of the corneal epithelium. Tr Am Ophthalmol Soc 62: 213–225

Cogan DG, Dickersin RG (1964) Nonsyphilitic interstitial keratitis with vestibulo-auditory symptoms. Arch Ophthalmol (Chicago) 71: 172–175

Obituary (1993) Washington Post, 11 September, B4

Obituary (1993) New York Times, 13 September

Sachs R (1967) Apraxie oculo-motrice Congenitale de Cogan. A propos de trois nouveaux cas dont deux dans la meme fratrie. Ann Oculist 200: 266–274

Vassella F, Lutschg J, Mumenthaler M (1972) Cogan's congenital ocular motor apraxia in two successive generations. Dev Med Child Neurol 14: 788–796

# COOLEY, Thomas B.

*(1871–1945)*

*From:* Am J Dis Child (1946) 71: 78

*Copyright:* 1946 American Medical Association

C OOLEY anaemia, or thalassaemia major, is an autosomal recessive condition which is common in Asia and the Mediterranean regions. Chronic haemolysis, stunted growth, skeletal changes and progressive hepatosplenomegaly are the major manifestations; death often occurs before puberty.

# Biography

C OOLEY was a paediatrician in the USA during the first half of the twentieth century; he made significant academic contributions in the field of haematological disorders of children.

Thomas Benton Cooley was born in 1871 at Ann Arbor, where his father was a justice of the Supreme Court and first dean of the law school of the University of Michigan. Cooley received a broad education and became fluent in several languages. He qualified in medicine at the University of Michigan in 1895 and subsequently interned at the Boston City Hospital. In 1905, after further postgraduate training in Germany and Ann Arbor, Cooley commenced practice in Detroit, where he was the first established paediatrician.

Cooley was active in community paediatrics as medical director of the Babies Milk Fund and collaborated with like-minded colleagues in the Association for the Study and Prevention of Infant Mortality. This group had considerable success in reducing infant mortality from diarrhoeal disease. During World War I Cooley served in France as assistant chief of the Children's Bureau of the American Red Cross; these activities were rewarded with the bestowal of the Cross of the Legion of Honour.

Cooley became chief of medical services at the Michigan Children's Hospital in 1921, and professor of paediatrics at Wayne University College of Medicine in 1936. He had sound judgement and a clear mind, attributes which he put to good use when he served as president of the American Paediatric Society and the American Academy of Paediatrics.

Cooley's academic contributions were made in the field of paediatric haematology. These included sections in standard textbooks, and the description of the condition which bears his eponym. Despite his great abilities, he wrote surprisingly little; he was a man of thoughts rather than actions.

Cooley was tall, with an austere aristocratic bearing and exquisite manners. Some regarded him as supercilious, whilst others interpreted his reticence as modesty. Although he was respected by his colleagues for his intellectual gifts, few knew him well. It is said that Cooley often speculated that he might have been a better lawyer than paediatrician.

A perspective of Cooley's personality can be gleaned from a statement made by Zuelzer in 1957: "Though an extremely able clinician and astute observer, he was more interested in the theoretical ramifications of the case at hand than in the individual patient. Rounds on his service at the Children's Hospital were occasions for brilliant and stimulating dissertations on a few cases selected for the challenge they offered to an analytical mind, or else for a review of urgent decisions which simply could not be put off any longer. The systematic coverage of the service, the detailed instruction of his resident staff failed to interest him."

At the time of his retirement in 1941, Cooley received emeritus status and an honorary doctorate of science. He died in 1945, being survived by a son and a daughter. The former followed family tradition and became a lawyer.

# Nomenclature

I N 1925 Cooley and his colleague, Lee, presented details of five children with splenomegaly, anaemia and unusual bone changes at a meeting of the American Pediatric Society. At that time, the anaemias of childhood were grouped together as "van Jaksch's anaemia"; in 1927, in a critical review, Cooley established the autonomous status of his patients' condition, which he subsequently termed "erythroblastic anaemia". When the geographical distribution of the disorder became apparent, the terms "Mediterranean anaemia" and "thalassaemia" (meaning "sea water in the blood") were introduced by Whipple and Bradford (1936), in conjunction with Cooley's name.

Dameshek (1940) used the eponym in a paper entitled "Target cell anaemia: anerythroblastic type of Cooley's erythroblastic anaemia". Other designations which have been employed are "Cooley-Lee syndrome", "chronic familial erythema" and "hereditary leptocytosis".

It is now known that the major and minor forms of thalassaemia represent the homozygous and heterozygous states of the condition; Cooley's anaemia corresponds to the former. In addition to thalassaemia, Cooley's name is also sometimes attached to an X-linked form of hypochromic anaemia, which he described shortly before his death. This condition, which has the alternative title "familial scleroblastic anaemia", is quite distinct from thalassaemia.

Thalassaemia confers relative immunity to malaria and the faulty gene is widely distributed in malarial regions. In high incidence communities the presence of the disorder has stimulated the development of techniques for antenatal diagnosis by foetal blood sampling. It has also served as a prototype for the understanding of the molecular pathology in a great many genetic conditions.

# References


Cooley TB (1927) Von Jaksch's anemia. Am J Dis Child 33: 786

Cooley TB (1945) A severe type of hereditary anemia with elliptocytosis: interesting sequence of splenectomy. Am J Med Sci 209: 561–568

Cooley TB, Lee P (1925) Series of cases of splenomegaly in children with anemia and peculiar bone change. Tr Am Pediat Soc 37: 29

Dameshek W (1940) "Target cell" anemia. Anerythroblastic type of Cooley's erythroblastic anemia. Am J Med Sc 200: 445–454

Obituary (1946) Thomas Benton Cooley, MD 1871–1945. Am J Dis Child 71: 77–79

Whipple GH, Bradford WL (1936) Mediterranean disease-thalassaemia (erythroblastic anemia of Cooley) associated with pigment abnormalities simulating haemachromatosis. J Pediatr 9: 279

Zuelzer WW (1957) Thomas B Cooley (1871–1945). *Pediatric Profiles*. Borden S Veeder (ed). The CV Mosby Co.

# CUSHING, Harvey Williams

## (1869–1939)

*From*: Fulton J.F. (1953) In: Webb Haymaker (ed) The Founders of Neurology

*Courtesy*: Charles C. Thomas, Springfield, Illinois
Portrait from: A bibliography of the writings of Harvey Cushing, prepared by the Harvey Cushing Society

C USHING syndrome is the result of excessive activity of the adrenal cortex, usually due to pituitary dysfunction, and manifests with obesity, a plethoric facies, hirsutism, hypertension and osteoporosis. The condition is predominantly non-genetic but uncommon familial forms have been documented.

# Biography

C USHING was an eminent North American surgeon. He is regarded as the founder of the speciality of neurosurgery.

Harvey Williams Cushing was born in Cleveland, Ohio on 8 April 1869, where his father, Henry Kirke Cushing, was a professor of obstetrics, gynaecology and medical jurisprudence. The family were of British stock and their progenitors had emigrated in 1638 to Bay Colony, Massachusetts from Norfolk, England. They had subsequently settled in Ohio where members of 2 generations had also been in the medical profession. Cushing was the youngest of 10 siblings and he was brought up in an orderly, disciplined home atmosphere.

Cushing attended Yale College and Harvard Medical School; before enrolling for the latter, he was directed by his father to abstain from smoking, drinking, boating, baseball and other forms of intemperance (doubtless many medical colleagues received similar admonitions from their fathers!). While still a medical student the death of a patient during a surgical procedure prompted him to design the first anaesthetic record chart, which was widely used. He also introduced the sphygmomanometer into the USA and blood pressure readings were routinely recorded on his chart. Cushing obtained his medical qualification in 1895 and after an internship at the Massachusetts General Hospital, he trained in general surgery under Halsted at the Johns Hopkins Hospital, Baltimore. During this period he also came under the influence of Welch, from whom he gained insights into medical history and education, and Osler, who taught him clinical neurology. Cushing's reputation was firmly established in 1910 when he successfully excised a meningioma from the brain of General Leonard Wood, the Chief of Staff of the US Army. Two years later Cushing was called to the chair of surgery at Harvard, where he was involved in the design of the Peter Bent Brigham Hospital; he spent the next 20 years at this hospital, prior to his retirement in 1932.

When World War I broke out Cushing took a surgical team from Harvard to the battlefields in France. In the later stages of the war he was attached to the Royal Army Medical Corps and received decorations for gallantry and distinguished service. Cushing contracted influenza during the pandemic of 1918 and thereafter suffered from chronic disability, which prevented him from taking more than a few steps at a time (this condition has been variously ascribed to arterial occlusive disease and peripheral neuropathy). Despite this handicap Cushing continued with his professional activities, made numerous contributions to neuroendocrinology and became the doyen of neurosurgery.

Cushing was a prolific medical author and his works ranged from a monograph on pituitary tumours published in 1912, to a classical review of meningioma, which appeared in 1938. His interest in medical history led him to write a 2-volume *Life of Sir William Osler* which took five years to complete and for which he received a Pulitzer Prize in literature in 1926.

Cushing was slim, dynamic, disciplined and meticulous. He was intolerant of inefficiency and demanded exemplary standards from his staff. He dressed fastidiously and abstained from alcohol, although he smoked excessively. Behind his cold, austere exterior lay great personal charm and he was admired by his peers. Cushing retired from his chair in 1932 and was appointed Sterling professor of neurology at Yale and consultant neurologist at New Haven Hospital. In 1937 he became professor emeritus and director of studies of medical history at Yale. By this time he had been a world figure for almost three decades, and he had received numerous medals, accolades, distinctions and honorary fellowships.

Cushing died of coronary heart disease on 7 October 1939 at the age of 70 years. His biographer, Elizabeth Thomson, stated "His death was not the end. Harvey Cushing, like a truly great teacher, had merely turned over his work to his pupils in clinics and operating rooms the world over".

# Nomenclature

C USHING had a life-long interest in the physiology of the pituitary gland and in 1932 he summarised his observations concerning the clinical manifestations of basophil adenomata. This disorder, which is mediated via the adrenal cortex, came to be known as the Cushing syndrome. The syndrome is aetiologically heterogeneous and usually non-genetic. Nevertheless, affected siblings with a possible autosomal recessive form of Cushing syndrome associated with adrenal cortical nodular dysplasia have been documented by Arce et al. (1978). Recessive inheritance was also postulated by Salti and Mufarrij (1981) and by Schweizer-Cagianut et al. (1980). Cushing syndrome may also occur as a component of the multiple endocrine neoplasia syndrome, but the eponym "Sipple" is more appropriately applied to this disorder (*see* p. 220).

Although Cushing's eponym is usually used in the context of adrenal dysfunction, it is also applied to an uncommon autosomal dominant disorder in which the proximal interphalangeal joints of the fingers are absent. Cushing encountered a woman with a brain tumour and rigid fingers at the Johns Hopkins Hospital in 1906 and during the next 10 years, in association with Professor Samuel Crowe, the head of Otolaryngology, he accumulated information concerning family members with the digital abnormalities. In his publication in which he documented the condition, Cushing described an autosomal dominant mode of transmission through 7 generations and introduced the term "symphalangism". The affected family was restudied 50 years later by Strasburger and colleagues and in a publication in 1965, the occasional syndromic component of conductive deafness was mentioned. The condition was traced back to William Brown, a Scottish immigrant who arrived in Virginia in the 1740s.

# References


Arce B, Licea M, Hung S, Padron R (1978) Familial Cushing's syndrome. Acta Endocr 87: 139–147

Cushing H (1916) Hereditary anchylosis of proximal phalangeal joints (symphalangism). Genetics 1: 90–106

Cushing HW (1932) The basophil adenomas of the pituitary body and their clinical manifestations (pituitary basophilism). Bull Johns Hopkins Hosp 50: 137–195

Schweizer-Cagianut M, Froesch ER, Hedinger C (1980) Familial Cushing's syndrome with primary adrenocortical microadenomatosis (primary adreno-cortical nodular dysplasia). Acta Endocr 94: 529–535

Salti IS, Mufarrij IS (1981) Familial Cushing disease. Am J Med Genet 8: 91–94

Steiner AL, Goodman AD, Powers SR Jr (1968) Study of a kindred with pheochromocytoma, medullary thyroid carcinoma, hyperparathyroidism and Cushing's disease: multiple endocrine neoplasia, type II. Medicine 47: 371–409

Strasburger AK, Hawkins MR, Eldridge R, Hargrave RL, McKusick VA (1965) Symphalangism: genetic and clinical aspects. Bull Johns Hopkins Hosp 117: 108–127

# DANDY, Walter Edward

*(1886–1946)*

*Courtesy*: Professor A.E. Walker, New Mexico

D ANDY-WALKER syndrome is a developmental defect of the foramina of Luschka and Magendie, which presents as hydrocephalus in infancy. Aetiological heterogeneity is likely and there is evidence for autosomal recessive inheritance in some instances.

# Biography

D ANDY was a distinguished neurosurgeon at the Johns Hopkins Hospital, Baltimore. He was a pioneer in his field and introduced many innovative operative techniques.

Walter Dandy was born in Sedalia, Missouri in 1886 shortly after his parents had emigrated from England. He obtained a Bachelor of Arts degree at the University of Missouri in 1907, declined a Rhodes Scholarship and proceeded to the Johns Hopkins Hospital School of Medicine, where he qualified in 1910. Dandy then spent the next eight years in resident surgical posts at the Johns Hopkins Hospital, before entering private practice in Baltimore.

It became evident at an early stage that Dandy had special talents and in 1913, at the age of 27 years and only 3 years after qualification, he published a classical account of the pathogenesis and management of hydrocephalus. At the time, his mentor, Halsted, commented "Dandy will never do anything equal to this again. Few men make more than one great contribution to medicine." In fact, this was only the beginning, and Dandy went on to make further notable discoveries.

In 1918 Dandy developed the technique of air encephalography and in 1925 he devised a new operative procedure for the treatment of trigeminal and glossopharyngeal neuralgia. Three years later he successfully treated Ménière disease by partial division of the acoustic nerve and in the same year he documented the operative management of ruptured intervertebral discs. Dandy also pioneered operations for intracranial aneurysms, acoustic neuromata and frontal lobe tumours.

Dandy had numerous positive attributes, including clarity and independence of thought, diligence, manual dexterity and the courage of his convictions. He was a prodigious author and amongst his many publications he wrote an outstanding monograph on the treatment of orbital tumours.

In his spare time Dandy played golf, tennis and bridge; he was also an expert on the history of the American Civil War. He had a happy home life and his son followed him into medicine.

In 1946 Dandy's sixtieth birthday was to have been marked by the dedication of an issue of the journal *Surgery*, which contained a set of articles written by his former pupils. He died shortly before publication and the contribution served to honour his memory. Dandy's outstanding qualities were summarised in an editorial in the *Baltimore Sun* at the time of his death: "He has the imaginative genius to conceive of new and startling techniques, courage to try them and skill – superlative skill – to make them successful".

# Nomenclature

I N 1914 Walter E. Dandy and his paediatric colleague, Kenneth Blackfan, published details of a female infant who developed hydrocephalus following a meningitic febrile illness at the age of 9 months. Investigations indicated that there was a block between the ventricles and the subarachnoid space. The child died a month later and at autopsy, a large, thin-walled cyst was found to fill the posterior fossa; the ventricular system was dilated and

foramina of Magendie and Luschka were occluded. The authors presented an elegant drawing of the post-mortem appearances of brain sections and attributed the internal hydrocephalus to congenital atresia of the foramina. Dandy (1921) subsequently mentioned that he had encountered three additional examples of the same disorder.

Walker became involved with the condition in 1942, when he gave a presentation on anatomical abnormalities of the ventricular system of the brain and the conjoined eponym was introduced by Benda in 1954. The Dandy-Walker syndrome is now a well-recognised anomaly (*see* p. 185).

# References


Benda CE (1954) The Dandy-Walker syndrome or the so-called atresia of the foramen of Magendie. J Neuropath Exp Neurol 13: 14–39

Blalock A (1946) Walter Edward Dandy 1886–1946. Surgery 19: 577–579

Dandy WE, Blackfan K (1914) Internal hydrocephalus: an experimental, clinical and pathological study. Am J Dis Child 8: 406–482

Dandy WE (1921) Diagnosis and treatment of hydrocephalus due to occlusions of the foramine and Luschka and Magendie. Surg Gynecol Obstet 32: 112–124

Obituary (1946) Walter Edward Dandy. JAMA 130: 1257

# DARIER, Jean
## *(1856–1938)*

D ARIER disease or keratosis follicularis is an autosomal dominant dermatological disorder in which multiple brownish greasy papules are associated with brittle furrowed nails and lesions of the mucous membranes and cornea. Endocrine disturbance and diffuse pulmonary fibrosis are rare concomitants.

# Biography

D ARIER was a distinguished dermatologist in Paris at the turn of the twentieth century. He is regarded as one of the founders of the science of dermatopathology.

Darier's family originated in Dauphine, France but in 1685 his ancestors sought refuge in Switzerland in order to avoid persecution following the revocation of the Edict of Nantes. His parents moved to Budapest before his birth in 1856, and returned to Geneva when he was 8 years of age. By this time he was fluent in Hungarian, French and German.

Darier commenced his medical studies in Geneva at the age of 15 years and qualified in Paris in 1880. He took French nationality and remained in that country, dividing his time between hospital medicine and histopathology. He became chief of service in 1894 and confined his interest to dermatology after 1906. Darier was appointed to the Hôpital St Louis in 1910, where he remained until his retirement in 1921.

Darier was an outstanding clinical dermatologist, but his international reputation was based to a large extent upon his endeavors in the field of histopathology. In addition to the conditions which bear his name *(vide infra)*, he published seminal papers on atrophic lichen planus and cutaneous sarcoidosis, tuberculosis and leprosy. His *Précis de Dermatologie*, which appeared in 1909 was translated into English and German, and went into five editions. He was also the chief editor of the 8-volume *Nouvelle Pratique Dermatologique*, which was the standard text of its time. He received academic recognition by appointments as Professor of the College of France, Member of the Academy of Medicine, Commander of the Legion of Honour and President of the International League of Dermatology. He was awarded an honorary doctorate from the University of Budapest and was given a similar distinction by the University of Geneva on his 80th birthday.

Darier held the conviction that investigations of cutaneous histopathology were an essential part of the diagnostic process. He made extensive use of material of this type in his articles and lectures, and founded the Museum of Histology at the Hôpital St Louis, for the preservation of his histological slides.

Darier was an innovator in the use of radiotherapy, chemotherapy and vaccines, and his enthusiasm was evident in his lectures, which attracted large audiences. In his obituary he was described by Graham-Little as being: "of middle height and of slender figure, which he retained to the end, as well as an undiminished head of hair. He wore a small and well-kept beard. His face was dominated by a noble Roman nose, truly symbolic of his character in its Roman simplicity and thoroughness. He gave one throughout his life the impression of abundant energy, vivacity, and those indefinable personal qualities summed up in the word 'charm'; he cultivated a certain elegance in his dress, and was a collector of beautiful objects."

In 1921 Darier retired to his country estate in the village of Longpont, near Paris. He was mayor of the village for more than a decade and was much concerned with local issues. He also retained an interest in many branches of science and continued with his medical writings until the day of his death. His health deteriorated during the last 3 years of his life, due to series of heart attacks, and he died at Longpont in 1938.

# Nomenclature

D ARIER'S name is conventionally used as a designation for keratosis follicularis. It is also sometimes applied to pseudoxanthoma elasticum and erythema annulare centrifugum.

In 1889 Darier documented the clinical and dermatopathological features of a condition which he termed "follicular psorospermosis." He gave a detailed description of two male patients attending the Hôpital St Louis and commented that their clinical and histological features were so concordant that the disease warranted a separate place in the framework of cutaneous nosography. In the same year, J.W. White published an account of the disorder, using the designation "follicular ichthyosis." The term "keratosis follicularis" is now firmly established, with "Darier disease" as an alternative. The conjoined eponym Darier-White is still sometimes used, but in general, the name of White has faded into obscurity.

Darier was involved in the early description of the clinical and histological features of the condition which is now known as pseudo-xanthoma elasticum. Although his name is sometimes applied to that disorder, the descriptive designation or the eponym Groenblad-Strandberg is preferred. There is scope for further semantic confusion, as Darier's name is also occasionally used for erythema annulare centrifugum. The aetiology of this recurrent skin disorder is unknown, but it is apparently non-genetic.

# References

Civatte J (1979) Jean Darier: a memoir. Am J Dermatopathol 1: 57–60
Darier J (1889) De la psorospermose folliculaire végétante. Ann Dermatol Syph 10: 597–612
Darier J (1896) Pseudoxanthoma elasticum. Mschr Prakt Derm 23:609–617
Darier J (1916–17) De l'érythème annulaire centrifuge (érythème papulo-circiné migrateur et chronique) et de quelques éruptions analogues. Ann Derm Syph, Paris 6: 57–58
Graham-Little E (1938) Obituary. Br J Dermatol 50: 384–389
Mitchell JH (1960) Some French dermatologists I have known. Arch Dermatol 81: 962–968
White JC (1889) A case of keratosis (ichthyosis) follicularis. J Cutan Genitourin Dis 7: 201–209

# DAY, Richard Lawrence

*(1905–1989)*

*From*: Pediatric Research (1986) 20 (10): 1011
*Courtesy*: The Williams & Wilkins Co, Baltimore

RILEY-DAY syndrome (*see* Riley, p. 143).

# Biography

DAY was an outstanding paediatrician and medical researcher in the USA during the twentieth century. He made major contributions to the understanding of thermoregulation and kernicterus in the newborn.

Richard Lawrence Day was born in New York City in 1905 and spent his boyhood in Summit, New Jersey. He obtained his higher education at Harvard and graduated in medicine in 1933. He was interested in paediatrics from the onset of his career and after internship at the Babies Hospital, New York, he trained in that speciality at Columbia. In 1937 he joined the staff of Cornell's Hospital, New York, where he was involved in the investigation of thermoregulation in neonates. His studies, which were undertaken in collaboration with James Hardy, a physicist at the Russell Sage Institute, form the basis for current concepts concerning the prevention of heat loss in the newborn.

Day's interest in thermoregulation led to his posting to the US Army Climatic Research Laboratory during World War II. Here he was involved in the development of insulated gloves for use in arctic conditions, and he was able to demonstrate that the ratio of surface area to volume of a limb or digit was a crucial factor in heat exchange.

After the war Day returned to the Columbia-Presbyterian Medical Centre, New York where he undertook investigations into kernicterus of the newborn. By means of experiments on rats he was able to demonstrate that excess bilirubin was directly responsible for cerebral damage.

Day became professor of paediatrics at the New York Downstate Medical School in 1953 and he subsequently occupied a similar post at the University of Pittsburgh. In 1965 he became the medical director of Planned Parenthood International, and in 1968 he joined the department of paediatrics at the Mount Sinai School of Medicine, New York. In 1975, after his retirement, his contributions to paediatrics were acknowledged with the establishment of the Richard L. Day lectureship at the Babies Hospital, Columbia University and his receipt in 1986 of the Howland Award of the American Pediatric Society. On the occasion of the presentation of this Award, his biographer, William A. Silverman described Day as "an enthusiastic, exciting person with all sorts of unusual ideas."

Day's enthusiasm for research continued unabated after his retirement and he was involved in projects on subjects as diverse as acupuncture, aspiration of foreign bodies and the mechanics of oars used in rowing. Richard Day and his wife were well known for their informal hospitality to groups of visitors and young people and their large house was often full of guests. Day died on 15 June 1989, aged 84 years, at his home in Westbrook, Connecticut. He was survived by Ida, his wife of 53 years and by their three daughters.

# Nomenclature

IN 1949 Conrad M. Riley, Richard L. Day and 2 of their colleagues published a detailed account of 5 children seen over a 10-year period at the Columbia University Babies Hospital, New York. The authors entitled their paper "Central autonomic dysfunction with defective lacrimation" and emphasised that the affected children all had similar features and that they were all of Jewish stock.

The authors recognised the ethnic homogeneity of their patients but concluded that there was no positive evidence for any "racial influence".

Once a condition has been delineated it often happens that affected persons turn out to be far more numerous than previously supposed. This disorder was no exception and by 1952 Riley (*see* p. 143) was able to publish details of 33 patients, including the 5 who had been described in the original paper. Riley termed the condition "familial autonomic dysfunction", confirmed the Jewish predilection, pointed out that his series included 4 pairs of siblings and that there was a positive history in 7 additional families. He concluded that "the proportions are compatible with Mendelian recessive inheritance".

The eponymic term "Riley-Day" syndrome was introduced in 1964, when Yatsu and Zussman published autopsy details following the death at the age of 31 years of one of the patients described in the original article. Goldstein-Nieviazhski and Wallis retained the same format in a review of 27 cases in Israel, published in 1966. A dysautonomia centre was subsequently established at New York University and by 1982, 227 affected persons had been seen in this special clinic. The eponym "Riley-Day syndrome" is generally accepted, with "familial dysautonomia" and "hereditary sensory and autonomic neuropathy" being used as alternative descriptive designations.

# References

Goldstein-Nieviazhski C, Wallis K (1966) Riley-Day syndrome (familial dysautonomia): survey of 27 cases. Ann Paediatr 206: 188–194

Riley CM, Day RL, Greeley DM, Langford WS (1949) Central autonomic dysfunction with defective lacrimation: report of five cases. Pediatrics 3: 468–478

Riley CM (1952) Familial autonomic dysfunction. JAMA 149: 1532–1535

Silverman WA (1989) In Memoriam. Richard L Day. An avant courier in neonatal medicine. J Perinatology 9: 244–245

Silverman WA (1986) Richard L. Day – The quintessential skeptical inquirer. Presentation of the Howland Award 1986. Pediatr Res 20(10): 1009–1012

Yatsu F, Zussman W (1964) Familial dysautonomia (Riley-Day syndrome). Case report with post-mortem findings of a patient at age 31. Arch Neurol 10: 459–463

# DIEKER, Hans Jochen

*(1941–1973)*

*Courtesy:* Professor H.-R. Wiedemann, Kiel

**M**ILLER-DIEKER syndrome is a lethal autosomal recessive disorder in which mental retardation, failure to thrive, a characteristic facies and malformation of internal organs are associated with lissencephaly (i.e. a smooth brain which lacks the normal convolutions).

# Biography

**D**IEKER was a German psychiatrist in the latter half of the twentieth century. His tragic death curtailed a potentially brilliant career in psychiatric genetics.

Hans Jochen Dieker was born on 18 April 1941 in Heidelberg, Germany. He graduated in medicine with honours from the University of Heidelberg, having also studied in Tubingen and Vienna. Following his internship Dieker trained in internal medicine in Mannheim and in psychiatry in Munich. In 1967 he joined Professor John Opitz for a year's post-doctoral fellowship in medical and paediatric genetics at the University of Wisconsin, Madison, USA. He took a special interest in malformations, and was involved in the delineation of several disorders, including the acrorenal polytopic field defect and the lissencephaly syndrome which bears his name.

In addition to his medical talents, Dieker was a classicist and a gifted musician, excelling on the flute. He died on 22 February 1973 in Munich, Germany, where he had been working at the Max Planck Institute on the genetic causes of mental retardation.

# Nomenclature

**I**N 1963, while undertaking a fellowship in neuropathology at Boston Children's Hospital, Miller documented 2 siblings with a lethal syndrome comprising microcephaly, a pinched facial appearance and temporal hollowing of the cranium. Autopsy revealed malformations of the kidneys, gut and heart, and the brain lacked convolutions or gyri. This latter abnormality, which is termed "lissencephaly" is the consequence of incomplete embryonic neuronal migration.

Dieker investigated a number of malformation syndromes during his post-doctoral fellowship at the University of Wisconsin and in 1969 he was senior author of a review article on the lissencephaly syndrome. This paper contained a detailed description of 2 affected brothers and a cousin with the agyria disorder, which had previously been delineated by Miller. Dieker and his colleagues mentioned that the occurrence of affected siblings with normal parents was suggestive of autosomal recessive inheritance.

In 1983 Dobyns and colleagues observed that lissencephaly could be associated with an abnormality of chromosome 17 and used the conjoined eponym "Miller-Dieker" in the title of their paper. Numerous articles followed and the conjoined eponym became firmly established.

The term "lissencephaly" is now used as a description of an anatomical abnormality, rather than the name of a specific disorder and numerical designations are applied to sub-types, on a basis of histopathological changes in the affected brain. The Miller-Dieker syndrome is regarded as type I lissencephaly; type II embraces the Walker-Warburg syndrome, and type III occurs in the Neu-Laxova syndrome. The molecular basis of the Miller-Dieker syndrome has now been elucidated and the gene "lissencephaly I" which is situated on chromosome 17 was cloned by Reiner and colleagues in 1993.

# References

Dieker H, Edwards RH, Zu Rhein G, Chou SM, Hartman HA, Opitz JM (1969) *The lissencephaly syndrome. The Clinical Delineation of Birth Defects II. Malformation Syndromes.* National Foundation-March of Dimes, New York, pp. 53–64

Dobyns WB, Stratton RF, Parke JT, Greenberg F, Nussbaum RL, Ledbetter DH (1983) The Miller-Dieker syndrome: lissencephaly and monosomy 17p. J Pediat 102: 552–558

Dobyns WB, Stratton RF, Greenberg F (1984) Syndromes with lissencephaly. I: Miller-Dieker and Norman-Roberts syndromes and isolated lissencephaly. Am J Med Genet 18: 509–526

Miller JQ (1963) Lissencephaly in 2 siblings. Neurology 13: 841–850

Reiner O, Carrozzo R, Shen Y, Wehnert M, Faustinella F, Dobyns WB, Caskey CT, Ledbetter DH (1993) Isolation of a Miller-Dieker lissencephaly gene containing G protein beta-subunit-like repeats. Nature 364: 717–721

# FALLS, Harold Francis
## *(b. 1909)*

*Courtesy*: Dr Richard A. Lewis, Houston, Texas

N ETTLESHIP-FALLS syndrome, or Ocular Albinism type I, is characterized by nystagmus, visual impairment and depigmentation of the retina. The condition is X-linked and female heterozygotes have patchy changes in the fundus of the eye (*see* Nettleship, p. 119).

# Biography

F ALLS enjoyed a long and fruitful career as an ophthalmologist and geneticist at the University of Michigan, USA.

In his autobiography, published in 1992 Falls stated "I was born in Winchester, Indiana on November 25, 1909 of rather mixed ancestry: On the paternal side Scots-Irish and on the maternal Dutch-English. When I was 3 years old my mother died of childbirth complications 6 months after delivering my sister. I went to live with my paternal grandmother and her second spouse, a farmer. A period of 5 pleasant and instructive years passed before my grandmother died of complications after being gored by an aroused cow. Grandmother was a Friend and belonged to a branch of the Quaker sect; she imbued me with the tenets of that religion. I was never to smoke, drink, or swear."

After the death of his grandmother, Falls moved with his father and stepmother to Detroit. He was educated at the Western High School in that city and then attended the University of Michigan, where he had considerable success, both academically and on the baseball field. After obtaining his medical qualification in 1936 and a higher degree and certification in ophthalmology in 1940, Falls was appointed to the staffs of the University of Michigan Hospital and the St Josephs Mercy Hospital, Ann Arbor. The Human Heredity Clinic at the University of Michigan, originally based in the undergraduate school's Department of Zoology, was directed by Falls during World War II. After the war the University recruited a rising scientist from the military, James Neel, to head the Department of Human Genetics. This was possibly the only time in history that a department of genetics grew out of a genetics unit headed by an ophthalmologist. Falls became Professor of Ophthalmology in 1959 and emeritus and honorary consultant status were conferred upon him at his retirement in 1975.

Falls lived an extremely busy life and in addition to his clinical and academic activities, he was a member of numerous medical university and public committees. These included the Michigan Chapter of the National Society for the Prevention of Blindness and the State Governor's Action Committee on Health Care. He also held offices in the American Academy of Ophthalmology and the American Society of Human Genetics and he played an important role in medical politics, through his involvement with the Michigan State Medical Society.

Falls was primarily a clinical researcher and he saw large numbers of persons with familial ophthalmological disorders. He had concern and compassion for his patients, whom he set at ease by his informal approach. Falls' research interests were focussed to a large extent upon genetic ophthalmology, and he is rightly regarded as a pioneer in this field. His first publication (familial retinoblastoma) appeared in 1940 and by 1976 he had been an author of more than 100 articles. He was one of the first to appreciate the importance of examining ostensibly normal relatives of affected persons and it was in this way that he recognised heterozygote manifestations in the condition which bears his eponym (vide infra).

Apart from his clinical duties Falls made a significant contribution to his hospital's postgraduate ophthalmology programme, teaching genetics, embryology and physiology. He also found time for regular perusal of the literature and had a reputation for being the most widely read member of the medical faculty. Falls had a prodigious memory and considerable skill as a writer; these talents served him well, both in his medical publications and in his committee reports.

Falls married Emeline Duckwitz and the couple had 3 children, Thomas, Harriette and Timothy. In his personal life, especially after retirement in 1975, Falls enjoyed his lakeside home, where his main pursuits were fly fishing for bass and playing golf with his wife. In 1994, at the age of 85 years Falls still maintained contact with his colleagues and took an active interest in ophthalmological matters.

# Nomenclature

O CULAR albinism was mentioned by Nettleship (1909) in a review of familial eye disorders. The patchy pigmentation of the retina in female heterozygotes with this X-linked disorder was depicted in an Atlas in 1942 by Alfred Vogt (*see* p. 175) but a report of these changes in the ophthalmological literature, written in 1951 by Falls achieved greater recognition and thereafter the eponym Nettleship-Falls came into use. In 1953 Francois and Deweer published drawings of the fundus of female heterozygotes and in 1962 Mary Lyon cited the mosaic pattern of pigmentation in ocular albinism as evidence to support her theory of random X-chromosomal inactivation. It is evident from Fall's article that he had recognised this process in the families of his patients with ocular albinism.

Forsius and Eriksson (1964) documented a large affected family on the Åland Islands in the Gulf of Bothnia, between Sweden and Finland, and Warburg further delineated the phenotype in the same year. In 1969 Waardenburg, together with Forsius and Eriksson suggested that the lack of fundal pigmentation in female heterozygotes in this family was indicative of heterogeneity and the Åland Island disorder was therefore given the designation "Ocular Albinism II" (Forsius-Eriksson) to distinguish it from Ocular Albinism-I, the Nettleship-Falls type. The presence of dermal macromelanosomes and visual pathway abnormalities in the former separates these conditions. At the molecular level the gene for Nettleship-Falls syndrome is in the Xp22.3 region, while that for the Forsius-Eriksson type is probably also on the short arm of the X-chromosome but closer to the centromere.

# References

Falls HF (1951) Sex-linked ocular albinism displaying typical fundal changes in the female heterozygote. Am J Ophthalmol 34: 41–50

Falls HF (1992) Scattered memorabilia: A living history. Autobiography of Harold F. Falls. Am J Med Genet 42: 153–155

Forsius H, Eriksson AW (1964) Ein neues Augensyndrome mit X-chromosomaler Transmission: eine Sippe mit Fundusalbinismus, Foveahypoplasie, Nystagmus, Myopie, Astigmatismus und Dyschromatopsie. Klin Mbl Augenheilk 144: 447–457

Francois J, Deweer JP (1953) Albinisme oculaire lei au sexe et alterations caracteristiques du fond d'oeil chez les femmes heterozygotes. Ophthalmologica 126: 209–221

Lyon MF (1962) Sex chromatin and gene action in the mammalian X-chromosome. Am J Hum Genet 14: 135–148

Nettleship E (1909) On some hereditary diseases of the eye. Trans Ophthalmol Societies of the UK 29: 57–198

O'Donnell FE, Green WR, McKusick VA, Forsius H, Eriksson AW (1980) Forsius-Eriksson syndrome: its relation to the Nettleship-Falls X-linked ocular albinism. Clin Genet 17: 403–408

Vogt A (1942) Die Iris: Albinismus solum bulbi. Atlas Spalt-Lampen-Mikroskopie 3: 846

Warburg M (1964) Ocular albinism and prolanopia in the same family. Acta Ophthomol 42: 444–451

# FØLLING, Asbjørn

## (1888–1972)

*Courtesy*: Professor Ivar Følling and Dr Ragna Følling Elgjo,
Norway

F ØLLING disease or phenylketonuria (PKU) is an autosomal recessive metabolic disorder in which defective activity of the enzyme phenylalanine hydroxylase leads to the development of mental retardation in early childhood.

# Biography

F ØLLING was a prominent medical biochemist at the University of Oslo, Norway, during the twentieth century. Many consider that he was unlucky not to be nominated for the Nobel Prize for his discovery of phenylketonuria.

Asbjørn Følling was born on 23 August 1888 at his parent's farm in Kvam, Norway, where he was the youngest of 6 children. His intellectual brilliance was apparent at an early age and he was sent to live with his married sister in Trondheim, so that he could benefit from schooling in that city. He qualified in chemistry in 1916 at the Technical College, Trondheim, and graduated in medicine at the University of Oslo in 1922. While studying medicine he supported himself by working in the physiology laboratory and in the summer vacation he shared the burdens on his parent's farm. In 1927 and 1930 he received Rockefeller research fellowships, which enabled him to visit the USA to study metabolic disorders at Harvard, Yale, the Mayo Clinic and the Johns Hopkins Hospital.

On his return to Norway, Følling proposed to follow a career in clinical medicine, but in 1932 he was called to the Chair of Nutritional Medicine at the University of Oslo. He missed contact with students and in 1934, when the Norwegian Veterinary College was opened, he accepted a post as professor of physiology and biochemistry. The years that followed were the happiest of his life, and he greatly enjoyed the prevailing synthesis of pioneering work, medicine and humanity. In 1953 he was called back to the University of Oslo as professor of medicine (biochemistry) and head of the Department and Institute of Clinical Biochemistry at the National Hospital (Rikshospitalet). When he retired in 1958, at the age of 70 years, Følling returned to the veterinary college, where he retained links throughout his retirement years.

Følling was a member of the Norwegian Academy of Science, and of the Societé de Chèmie Biologique, Paris. He was also an honorary member of the American Association on Mental Deficiency, USA, and of the Norwegian Medical Association. His discoveries relating to phenylketonuria were recognised by the Fridtjof Nansen Prize, the Jahre Medical Prize and the Joseph P. Kennedy, Jr. International Award, together with several awards from Norwegian and Danish civil and scientific associations.

Følling died in 1972 at the age of 84 years, after 4 years of disability, following a stroke. In an account of his life and career, his daughter, Dr Ragna Følling Elgjo made the following comments on her father's character and personality, "He was a very humble man. He did not like the slogan: Wisdom is power, but would rather say: Wisdom is humility. He used to say: We are all tools in the hands of the Lord, and he was thankful that he had been of some use.

He was a foresighted man – seeing the possibilities, and also the worries of the future. One example of this was the fact that he broke with tradition when he chose to go to technical school instead of staying in the agricultural milieu where he really belonged. He saw the possibilities in new technical knowledge. Very early, already around 1920, he was concerned about the misuse – or abuse – of the soil and resources of this planet, and of the concept of overpopulation.

He had a wonderful sense of humour and was a good storyteller. He was very well read, mostly being attracted by the philosophically orientated authors and the harmony of Greek art. Like many great men, he had a poetical soul and he also wrote some nice poems himself. And he saw the need for beauty in this harsh world."

# Nomenclature

I N 1934 the parents of 2 mentally retarded children asked Asbjørn Følling for advice. Both children had appeared to be normal at birth, but then gradually deteriorated. In addition to their mental retardation, they had decreased pigmentation of the skin and hair, with peculiarities of gait, stance and behaviour. Their father suffered from asthma, and thought that his attacks were precipitated by the peculiar mousy odour of his children's urine, which pervaded their room.

Følling examined the urine for ketones and found that an unusual green colour appeared on the addition of the appropriate reagent. This phenomenon was unknown at that time and he therefore decided to make an attempt to purify the compound that caused this reaction. After 6 weeks of work on large amounts of urine from the 2 children he managed to identify phenylpyruvic acid and phenylacetic acid in the urine. He assumed that there could be a relationship between the production of these urinary metabolites and the mental deficiency. He also found an increased concentration of phenylalanine in the patients' serum, and in 1934 he published his findings in the biochemical literature, naming the disease "phenylketonuria phenylpyruvica". Følling then tested the urine from several hundred patients with various forms of mental disturbance and identified 11 with the same reaction, of whom 4 were siblings. He concluded that he had found an inherited metabolic disorder which caused mental retardation, probably because of a toxic injury to brain cells caused by increased phenylalalanine levels in the blood. He proposed the name "imbecilitas phenylpyruvica". The significance of the discovery was, however, not fully understood at that time and almost 20 years elapsed before it was recognised as a crucial advance in research on mental retardation and inborn errors of metabolism.

In collaboration with the geneticist, Professor L. Mohr, Følling proved that the disorder was inherited as an autosomal recessive trait with a frequency in Norway of 1 per 20,000 births. Følling also succeeded in detecting heterozygous gene carriers by dietary loading of parents of affected children with an excess of phenylalanine. A comprehensive account of the discovery of phenylketonuria was published in 1994 by his son Ivar Følling.

# References


Følling A (1934) Uber Ausscheidung von Phenylbrenztraubensaure in den Harn als Stoffwechel anomalie in Verbindung mit Imbezilitat. Hoppe-Seylers Z Physiol Chem 227: 169–176

Følling A (1971) The Original Detection of Phenylketonuria. In: Bickel H, Hudson FP, Woolf LI (eds) *Phenylketonuria and Some Other Inborn Errors of Amino Acid Metabolism. Biochemistry, Genetics, Diagnosis, Therapy.* Georg Thieme Verlag, Stuttgart

Følling A, Closs K (1938) Uber das Vorkommen von 1-Phenylalanin in Harn und Blut bei Imbecillitas phenylpyrouvica. Hoppe-Zeylers Z Physiol Chem 254: 115

Følling A, Mohr OL, Ruud L (1944) Oligophrenia phenylpyrouvica. A recessive syndrome in man. Det Norske Videnskaps-Akademi, Skrifter. I Mat Nat Klasse No. 13: 1

Følling I (1994) The discovery of phenylketonuria. Acta Paediatr Suppl 407: 4–10.

Ringdal N (1992) *Kjepperi Varherres Hjul: en bok om Asbjørn Følling of sykdommen som fikk hans navn.* Scanbok

# GARDNER, Eldon

*(1909–1989)*

G ARDNER syndrome comprises colorectal adenomatosis, epidermal cysts, fibromas and osteomata, especially of the calvarium and jaws. There is a high risk of malignant degeneration of the adenomatous polyps during adulthood. Inheritance is autosomal dominant.

# Biography

G ARDNER was an eminent North American biologist and geneticist.

Eldon Gardner was born on 5 June 1909 in Logan, Utah. After receiving his education in that town he studied zoology at Utah State University, gaining his bachelors degree in zoological chemistry in 1934, and a masters degree in zoology in 1935 for a study of spermatogenesis. Gardner continued his academic career at the University of California, obtaining his doctorate in 1939 with a thesis on dominance modification and position effects in drosophila. During the period 1939–1946, Gardner held an appointment at the Salinas College, California, where he was involved in research into the production of artificial rubber. In 1946, Gardner became associate professor in the Department of Biology, University of Utah, Salt Lake City and in 1948 he returned to his old home in Logan, as a member of the University of Utah faculty. In 1962 he was appointed Dean of the new College of Science at Utah State University and in 1967 he became Dean of the School of Graduate Studies.

In the early phase of his career Gardner published extensively on drosophila and wrote textbooks on genetics and the history of biology. Later he wrote numerous articles on multiple exostoses and intestinal polyposis. During his distinguished career he received many academic awards, including an honorary doctorate of science from his own university. After his retirement in 1974 Gardner became emeritus professor of biology and continued the investigation of the condition which bears his name.

Gardner was active in community affairs and he held leadership positions in the Church of the Latter-day Saints. He was also a member of the local Chamber of Commerce and served as President of the Logan Rotary Club. He had a happy family life with 6 children and 24 grandchildren.

Gardner died from a heart attack on 1 February 1989, at the age of 80 years.

# Nomenclature

I N 1972 a volume in the Birth Defects Original Articles Series, on the topic of intestinal disorders, was dedicated to Eldon Gardner. During the conference from which these proceedings emanated Gardner gave a comprehensive account of the discovery of the syndrome which now bears his name. He recalled his early investigations in Utah and recounted that in 1947, while teaching genetics to an undergraduate class, he mentioned familial neoplasia. A student drew his attention to a large kindred with intestinal cancer and, following a home visit, Gardner initiated a comprehensive series of investigations, which continued for more than three decades.

Gardner recognised the propensity for malignant change in this inherited disorder and published his first account in 1951. Numerous articles followed, including the classical description of the syndromic association of sebaceous cysts, osteomata and intestinal polyposis, which was published in 1953. In 1958, Smith employed Gardner's name in the title of an article on polyposis and thereafter the eponym came into general use. It has now emerged that the Gardner syndrome is a comparatively common genetic entity and by 1994 more than 98 affected kindreds had been recorded in the Johns Hopkins Hospital colon polyposis registry.

In 1987 Bodmer and colleagues localised the gene for Gardner syndrome to the long arm of chromosome 5. Thereafter, with the development of molecular technology, it has become apparent that Gardner syndrome and familial intestinal polyposis without malignant potential result from different genes at the same chromosomal locus (i.e. allelic genes). Numerous different intragenic defects in separate families have been demonstrated and it is evident that there is considerable molecular heterogeneity in this disorder.

# References

Bodmer WF, Bailey CJ, Bodmer J, Bussey HJR, Ellis A, Gorman P, Lucibello FC, Murday VA, Rider SH, Scrambler P, Sheer D, Solomon E, Spurr NK (1987) Localization of the gene for familial adenomatous polyposis on chromosome 5. Nature 328: 614–616

Gardner EJ (1951) A genetic and clinical study of familial polyposis, a predisposing factor for carcinoma of the colon and rectum. Am J Hum Genet 3: 167–176

Gardner EJ, Richards R (1953) Multiple cutaneous and subcutaneous lesions occurring simultaneously with hereditary intestinal polyposis and osteomas. Am J Hum Genet 5: 139–147

Gardner EJ (1972) Discovery of the Gardner syndrome. The clinical delineation of Birth Defects XIII. GI tract, including liver and pancreas. Birth Defects Orig Art Ser VIII(2): 48–51

Smith WG (1958) Multiple polyposis, Gardner's syndrome and desmoid tumors. Dis Colon Rectum 1: 323–332

Woolf CM, Remondini DJ, Simmons JR (1989) Eldon J. Gardner (1909–89): In Memoriam. Am J Hum Genet 45: 471–473

# GLANZMANN, Eduard
## *(1887–1959)*

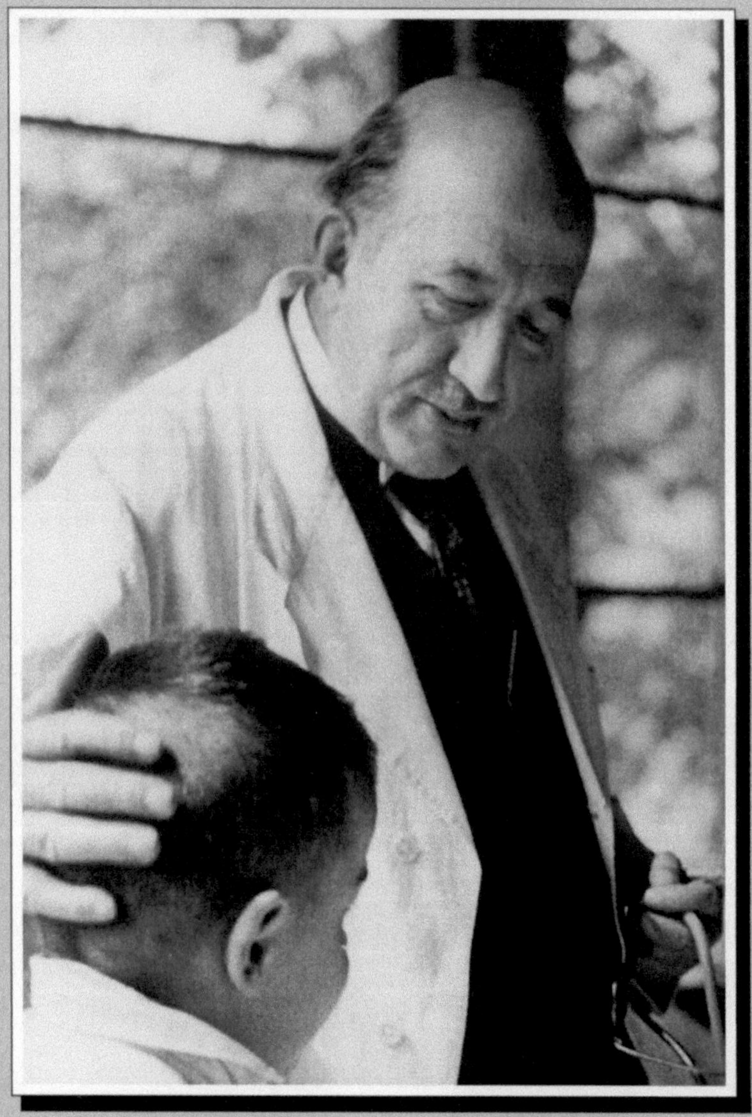

*Courtesy*: Professor D. Klein, Switzerland

G LANZMANN-NAEGELI syndrome, or thrombasthenia, is an autosomal recessive disorder characterised by anaemia, petechiae and a bleeding diathesis, due to platelet dysfunction and defective aggregation.

# Biography

G LANZMANN was professor of paediatrics at the University of Berne, Switzerland in the middle portion of the twentieth century. He made notable contributions in the field of paediatric haematology.

Eduard Glanzmann was born on 12 April 1887 in Lucerne, Switzerland, where his father was employed in the postal services. After attending school in his home town, he studied medicine at the Universities of Zurich, Berlin and Berne. Glanzmann obtained his medical qualification in 1914 and undertook postgraduate studies in Berlin, under Czerny, the "Pope of Paediatrics", before settling in Berne as a specialist in children's disorders. He maintained his academic links and in 1932 was appointed as professor of paediatrics at the University of Berne, in succession to his former teacher, Professor Stoos. Glanzmann occupied this chair until chronic ill-health led to his retirement in 1957.

Glanzmann was a busy clinician but he found time for applied research and he published extensively. His articles and books received world-wide recognition and his monograph *Introduction to Paediatrics* became a standard text. This classical work, which was written in German and translated into French and Spanish, went into several editions. It represented a review of current paediatric knowledge, ranging from clinical medicine through biochemistry to pathology, and information was presented in a lucid, lively style. The assistance of his wife, Dorothy, née Steinegger, whom he married in 1924, was crucial to these endeavours. Glanzmann's success led to the development of the Jenner Children's Hospital in Berne as an international postgraduate centre and he gained special satisfaction from the inception of a wing dedicated to infectious diseases. At one time he was called to a Chair at the University of Vienna; he declined, preferring to remain in his native land.

Glanzmann undertook research in the fields of infectious disease, vitamin therapy and, above all, haematology. He recognised the role of allergy in the pathogenesis of Henoch-Schonlein purpura and made original observations concerning the involvement of the endocrine glands in cystic fibrosis. He achieved wide international recognition and received accolades from paediatric and scientific societies in many countries; in 1952 the *Archives Françaises de Pédiatre* published a jubilee edition in his honour.

Glanzmann had a portly physique and a deep voice, which concealed a lively personality. He was a plain, approachable man, full of humanity, devotion, goodness and love for the children under his care, and he was held in high regard by his patients and colleagues alike.

In 1952, following an accident, Glanzmann's health began to fail. Nevertheless, he persisted with his clinical duties and continued with his publications and literature reviews. He became increasingly handicapped and was eventually confined to his bed; his mind remained clear, however, and he worked on a fresh edition of his monograph until shortly before his death in 1959 at the age of 72 years.

Glanzmann's life and character were summed up in his obituary "Untold patients and thankful mothers will mourn him. Enemies he had none. Swiss medical science become poorer by the loss of this worthy, kind man."

# Nomenclature

G LANZMANN'S interests in haematology were complemented by his studies of diet and infection, and he was particularly concerned with the influence of nutrition on childhood anaemia.

In 1918 Glanzmann published an account of a condition which he labelled "Hereditary Haemorrhagic Thrombasthenia". The disorder can be comparatively benign; the affected girl whom he described became symptomatic at the age of 7 years and survived until her death in Paris 78 years later. With the elucidation of the haematological phenotype, it was accepted that deficient clot retraction and abnormal platelet agglutination are the major features. Inheritance is usually autosomal recessive, although an uncommon dominant form has been documented and, as discussed by Alagille et al. (1964), further heterogeneity is likely. The condition is well recognised and the eponym "Glanzmann thrombasthenia" is widely used, although the additional eponym "Naegeli" is sometimes added. A form of the disorder in which platelet prothromboplastin is defective and the platelets are morphologically abnormal, is known as the Revol or Bernard-Soulier syndrome.

Glanzmann's name has been used in conjunction with that of Riniker in the title of congenital lymphoid hypoplasia, or Swiss agammaglobulinaemia. Although this genetically heterogeneous condition is well recognised, the eponym is now rarely employed.

# References


Alagille D, Josso F, Binet JL, Blin ML (1964) La dystrophie thrombocytaire hemorragipare. Discussion nosologique. Nouv Rev Franc Hemat 4: 755–790

Caen JP, Castaldi PA, Leclerc JC, Inceman S, Larrieu MJ, Probst M, Bernard J (1966) Congenital bleeding disorders with long bleeding time and normal platelet count. I. Glanzman's thrombasthenia (report of 15 patients). Am J Med 41: 4–26

Glanzmann E (1918) Hereditäre hämorrhagische Thrombasthenie. Ein Beitrag zur Pathologie der Blutplättchen. Jb Kinderh 88: 1–42, 113–141

Obituary (1959) Arch Fr Pediat 16(2): 288

Revol L (1950) Nouveau type de dysmorphie thrombocytaire. La diacyclothrombopathie. Lyn Méd 183: 213–218

# GOODMAN, Richard M

*(1932–1989)*

**G**OODMAN syndrome, or Tel-Hashomer campto-dactyly, comprises digital contractures, a distinctive facies, muscular hypoplasia and multiple skeletal abnormalities. Inheritance is autosomal recessive.

Goodman acrocephalopolysyndactyly (ACPS IV) is a rare autosomal recessive disorder in which congenital cardiac malformations and craniostenosis are associated with digital contractures.

# Biography

**G**OODMAN was a prominent medical geneticist in the USA and Israel. He was a prolific author and he is remembered for his classical monograph *Genetic Disorders Among the Jewish Peoples.*

Richard Goodman was born on 31 July 1932 in Cleveland, Ohio, USA. He received his medical education at Ohio State University and undertook residency training in internal medicine at Cook County Hospital, Chicago. The plethora of rare conditions which he encountered during this period kindled his interest in medical genetics and he subsequently held a fellowship with Victor McKusick at the Johns Hopkins Hospital, Baltimore and spent a year in Israel studying genetic disorders with the late Dr Chaim Sheba.

Goodman returned to the USA in 1964 as Head of the Division of Medical Genetics at Ohio State University. His wide experience of clinical genetics and his talents as an author found expression in a number of books, including *Genetic Disorders of Man, The Face in Genetic Disorders, An Atlas of the Face in Genetic Disorders* and *The Malformed Infant and Child: An Illustrated Guide.*

Goodman had a profound awareness of his religious background and in 1971 he moved with his family to Israel, where he became Professor of Human Genetics at the Tel-Aviv University, Sackler School of Medicine and the Chaim Sheba Medical Center at Tel-Hashomer. Goodman's knowledge of Jewish history and his appreciation of the opportunity for genetic research presented by the immigration of diverse populations to Israel, culminated in his classical monograph *Genetic Disorders Among the Jewish Peoples.* This work had its roots in the influence of the late Professor Chaim Sheba, who had initially excited Goodman's interest in this field. A second volume *Genetic Disorders Among Ashkenazi Jews* was co-authored with Arno Motulsky. Goodman was constantly alert for new genetic syndromes and he eventually published more than 130 articles, mainly in the field of dysmorphology.

Goodman was happily married to Audrey and their 3 children were a source of great joy to him. He immersed himself in Israeli culture and obtained fulfillment from the orthodox lifestyle which he followed. His home on the northern outskirts of Tel Aviv was close to archaeological sites, thus facilitating his hobby of collecting artifacts. The seashore was also nearby and he enjoyed walking on the beach – hence his book of poems *Children Along the Seashore.* As a young man Goodman was noted for his affability and sense of humour; by middle age he had become increasingly philosophical and was developing the attributes of a sage.

Goodman developed carcinoma of the colon in 1989 and died in the summer of that year at the age of 57 years. David Comings, his friend and biographer, stated "I am sure that he was upset with his illness not because of a concern with death but because he was angry with the loss of family and Israel, and the inability to finish his next book."

# Nomenclature

**W**ITH the immigration of people from diverse geographical backgrounds into Israel, Goodman had the opportunity to recognise and delineate several new genetic syndromes. In 1972, together with his colleagues at the Chaim Sheba Medical Center, he documented a brother and sister of Moroccan Jewish stock who had camptodactyly and muscular hypoplasia. In 1976 they recognised the same condition in a pair of sisters from a consanguineous Arab Bedouin kindred, and suggested the name "Tel-Hashomer camptodactyly syndrome". Further cases were reported from other parts of the world; the phenotype was expanded and autosomal recessive inheritance was confirmed. The designation "Tel-Hashomer" was usually employed in the titles of these articles and Goodman's name was often attached to the disorder.

Goodman and his colleagues maintained their interest in digital contractural syndromes and in 1979 they reported three siblings from a consanguineous Iranian Jewish family who had flexed fingers, polysyndactyly, congenital cardiac malformations, and craniostenosis. They suggested that this condition was a new autosomal recessive trait and they proposed the designation "acrocephalopolysyndactyly type IV". This disorder was listed in the 10th edition of McKusick's *Catalogue of Mendelian Inheritance in Man* and the term "Goodman syndrome" was used in the title. The syndromic identity of the Goodman syndrome was questioned in 1987 by Cohen and colleagues when they suggested that it represented part of the wider phenotypic spectrum of the well established Carpenter syndrome. This nosological issue remains unsettled.

# References

Goodman RM, Katznelson MB, Manor E (1972) Camptodactyly: occurrence in two new genetic syndromes and its relationship to other syndromes. J Med Genet 9: 203–212

Goodman RM, Katznelson M, Katznelson A (1976) Camptodactyly with muscular hypoplasia, skeletal dysplasia and abnormal palmar creases: Tel-Hashomer camptodactyly syndrome. J Med Genet 13: 136–141

Goodman RM, Sternberg M, Shem-Tow Y, Katznelson MB, Hertz M, Rotem Y (1979) Acrocephalopolysyndactyly type IV: a new genetic syndrome in 3 sibs. Clin Genet 15: 209–214

McKusick VA (1992) *Mendelian Inheritance in Man.* 10th Edn. Johns Hopkins University Press, Baltimore & London, p. 1181

Cohen DM, Green JG, Miller J, Gorlin RJ, Reed JA (1987) Acrocephalopolysyndactyly type II – Carpenter syndrome: clinical spectrum and an attempt at unification with Goodman and Summitt syndromes. Am J Med Genet 28: 311–324

Obituary (Comings DE) (1991) Am J Med Genet 38: 517

# GORDON, Hymie

*(1926–1995)*

G ORDON syndrome is an autosomal dominant disorder comprising camptodactyly, club foot and cleft palate.

# Biography

G ORDON was a South African physician and medical geneticist. The major part of his career was spent at the Mayo Clinic, USA.

Hymie Gordon was born in Mowbray, Cape Town, South Africa on 20 September 1926. He attended the Wynberg Boys High School and the University of Cape Town where he graduated with a BSc in Medical Sciences in 1946 before proceeding to a medical qualification in 1950 and a doctorate in 1958.

Gordon sub-specialized in cardiology, which he studied at the Royal Postgraduate Medical School in London. His main work there was on a classification of the "idiopathic" cardiomyopathies into hypertrophic, dilated and restricted types. He had planned to continue this work at the Johns Hopkins Hospital in Baltimore, USA, under the auspices of the cardiologist Victor A. McKusick; in a letter to the authors Gordon commented "by 1959 when I crossed the Atlantic, McKusick had mutated into a geneticist and I therefore became a Fellow in genetics".

On his return to the University of Cape Town in 1961 Gordon was appointed as a senior lecturer in medicine, and he established a comprehensive medicine group, in which he tried to introduce a multifactorial approach to medical practice, research and education by integrating genetic and ecological approaches. He also gave the first systematic course in genetics at Cape Town Medical School; he stated "this included almost all that was known about molecular genetics (in 30 minutes) and about human cytogenetics (in 35 minutes)." Gordon's major research activities included the analysis of biochemical genetic markers, especially red-cell enzyme polymorphisms, in Cape Town's different ethnic groups, a survey of the ecology of hypertension in these populations and field studies of several Mendelian disorders, notably lipid proteinosis.

In 1969 Gordon moved to the Mayo Clinic, where he established a department of medical genetics, which he chaired until he retired from clinical practice in 1990. He was granted emeritus status and thereafter studied and taught history, including a course entitled "History for Physicians".

Gordon was a bachelor and retained close links with his relatives in Cape Town. He died suddenly on 2 February 1995 while on holiday in that city.

# Nomenclature

I N 1969 Gordon, together with his Groote Schuur Hospital colleagues, David Davies, a plastic surgeon, and Mervyn Berman, a chemical pathologist, investigated a young man with a cleft palate plus hand and foot contractures. Family studies revealed that 5 relatives in 3 generations of the Cape Town family were affected and the researchers proposed that the condition was an autosomal dominant trait.

Another family with 3 affected members was reported by Higgins et al. in 1972 and a further large kindred was documented in 1979 by Halal and Fraser. The title of the latter article contained the eponym "Gordon syndrome" in addition to a descriptive designation. In a large scale review of the arthrogryposes Hall et al. (1982) termed the Gordon syndrome "distal arthrogryposis type IIA" and they suggested that early cases had been documented by Moldenhauer (1964) and Krieger and Espiritu (1972).

The gene for a form of distal arthrogryposis has been localized to chromosome 9 and it is possible that the mutation underlying the Gordon syndrome is situated at this site. By 1995 the original Cape Town family has expanded to 5 affected generations, and arrangements for Gordon to review his old patients were forestalled by his sudden death in the week prior to the proposed reappraisal.

# References

Gordon H, Davies D, Berman MM (1969) Camptodactyly, cleft palate and club foot: a syndrome showing the autosomal dominant pattern of inheritance. J Med Genet 6: 266–274

Halal F, Fraser FC (1979) Camptodactyly, cleft palate and club foot (the Gordon syndrome): a report of a large pedigree. J Med Genet 16: 149–150

Hall JG, Reed SD, Greene G (1982) The distal arthrogryposes: delineation of new entities – review and nosologic discussion. Am J Med Genet 11: 185–239

Higgins JV, Hackel E, Kapur S (1972) A second family with cleft palate, club feet and camptodactyly. (Abstract). Am J Hum Genet 24: 58A

Krieger I, Espiritu CE (1972) Arthrogryposis multiplex congenita and the Turner phenotype. Am J Dis Child 123: 141–144

Moldenhauer E (1964) Zur Klinik des Nielson-Syndroms. Derm Wschr 150: 594–601

Obituary (1995) Cape Times, 7 February p. 7

Obituary (1995) S Afr Med J 85(8): 800–801

# GOWERS, Sir William Richard

*(1845–1915)*

*From*: Duke-Elder S. Sir (1971) System of Ophthalmology XII
*Courtesy*: Mosby-Year Book, Inc, St. Louis

G OWERS muscular dystrophy is an autosomal dominant, slowly progressive disorder in which weakness and wasting of the small muscles of the hands and feet commences in adulthood.

# Biography

G OWERS was a founder of British neurology, and practised in London in the latter years of the nineteenth century. He is remembered for his classical medical texts and for his artistic skills.

William Richard Gowers was born in London on 20 March 1845. He attended Christ's Church College School, Oxford and at the age of 16 years he was apprenticed to a country practitioner at Coggeshall, Essex. Thereafter he studied medicine at University College, London, qualifying in 1869. His intellectual brilliance was already apparent at this stage and he was awarded gold medals for his academic efforts in several separate fields. Gowers proceeded to a doctorate in 1870, obtaining this degree with distinction.

With the influence and guidance of his patron and mentor, Sir William Jenner, Gowers was appointed to the National Hospital for the Paralysed and Epileptic. He was also on the staff of the University College Hospital and Medical School, ultimately becoming Professor of Clinical Medicine. He was admitted to the Fellowship of the Royal College of Physicians in 1879 and delivered the prestigious Goulstonian lecture in 1880, choosing epilepsy as his subject. Election to the Fellowship of the Royal Society followed in 1887.

Gowers was a scientific innovator, with wide ranging interests. In 1878 he constructed a haemoglobinometer, which remained in general use for the next 20 years. He was also an early proponent of ophthalmoscopy and in 1879 he published *A Manual and Atlas of Medical Ophthalmoscopy*. Gowers had a long-standing interest in neuroanatomy, which found expression in 1880 in the publication of his book *The Diagnosis of Diseases of the Spinal Cord*. At this stage of his life, Gowers' output was prodigious; *Lectures on the Diagnosis of Diseases of the Brain* appeared in 1885, followed by *A Manual of Disease of the Nervous System* in 1886. This latter work, which was a synthesis of existing knowledge together with his own vast experience, represents his magnum opus. It achieved world-wide recognition as a standard text and went into several editions. Gowers' systematic approach to the delineation and elucidation of neurological disease rivalled that of the great Charcot and he ranks amongst the founders of his speciality. His scientific contributions received due recognition in 1897, when he was knighted during the Diamond Jubilee of Queen Victoria. His status and international reputation resulted in the award of numerous honorary Doctorates, Fellowships and academic society memberships.

Gowers had an incisive mind and a vast memory; in the early years of his career he was impatient with persons of lesser intellect. Like many colleagues before and since, he mellowed with age. Gowers was methodical and precise in his approach to disease and research and he collected details of more than 20,000 neurological cases, which he recorded in shorthand. In addition to his medical skills, Gowers was a talented artist and he was able to illustrate his own books. He also regularly exhibited his paintings and etchings at the Royal Academy.

Gowers married Mary Baines of Leeds in 1875 and the couple had 2 sons and 2 daughters. He died in London on 4 May 1915, after a protracted illness.

# Nomenclature

T HE fact that Gowers' name ends with the letter "s" has been a source of long-standing confusion. As with Perthes disease, there is a wide misconception that the surname, when used eponymously, represents the possessive form.

Gowers had a special interest in neuromuscular disorders and he was largely responsible for drawing attention to the pseudohypertrophic form of muscular dystrophy, which had been delineated by Duchenne. In a series of 5 clinical lectures on this subject, delivered to students of University College, London, at the National Hospital for the Paralysed and Epileptic, he outlined the characteristic way in which affected children rise to their feet from a sitting position by pushing the trunk upwards through pressure exerted by their hands on the thighs. Since that time, the term "Gowers sign" has been applied to this manoeuvre.

Gowers' eponym is used as a designation for distal myopathy. He described this condition in 1902 giving details of a young affected male, under the title *A lecture on myopathy of a distal form*. Further reports were scanty until Welander (1951) documented 249 affected persons in 72 families in Sweden and emphasized the dominant mode of inheritance, late onset and slow progression. Other Swedish families were subsequently described by Edstrom (1975) and it is evident that, although rare elsewhere, the condition is comparatively common in Scandinavia.

The term "distal myopathy" is firmly established but there is now a tendency to omit Gowers' eponym. This is largely the result of a paper by Markesbery et al. (1977), in which it was suggested that the original patient described by Gowers might have had dystrophia myotonica rather than distal myopathy. Be that as it may, Gowers was an academic giant in his time, and he richly deserves eponymous immortality!

# References


Critchley M (1949) *Sir William Gowers. A Biographical Appreciation*. William Heinemann, London

Edstrom L (1975) Histochemical and histopathological changes in skeletal muscle in late onset hereditary distal myopathy (Welander). J Neurol Sci 26: 147–157

Gowers WR (1902) Myopathy of a distal form. Br Med J 2: 89–92

Markesbery WR, Griggs RC, Herr B (1977) Distal myopathy – electron microscopic and histochemical studies. Neurology 27: 727–735

Obituary (1915) Lancet 1: 1055–1056

Obituary (1915) Br Med J 1: 828–830

Welander L (1951) Myopathia distalis tarda hereditaria. Acta Med Scand 141 (Suppl 265): 1–124

# GREBE, Hans
*(b. 1913)*

G REBE chondrodysplasia is an autosomal recessive skeletal disorder characterised by severe shortening of the digits and distal portions of the limbs.

# Biography

G REBE is a German physician. He was a notable athlete and a founder of the speciality of sports medicine.

Hans Grebe was born on 25 August 1913 in Frankfurt, Germany, where his father was a rector. During his boyhood he enjoyed sporting activities and was an active member of the "jung wandervogels". He studied medicine and physical culture concurrently at the Universities of Berlin and Frankfurt, obtaining the medical faculty prize and graduating *summa cum laude* in 1937.

Grebe was wounded on 3 occasions while serving as a medical officer in World War II, and received the Iron Cross, 1st class. He left the military in 1942 to become director of Human Genetics at the Kaiser Wilhelm Institute of Anthropology, Human Genetics and Eugenics in Berlin. In 1944 he became professor and director of the Institute for Genetic Biology and Eugenics at the University of Rostock. After the war Grebe settled in private practice in Frankenberg, Eder, and in 1953 he obtained a teaching appointment at the University of Marburg, which he held until 1973.

Grebe was talented, active, optimistic and enterprising. These qualities are reflected in his medical publications; he was a prodigious author and wrote more than 350 medical articles, together with 3 monographs and 15 textbooks, mainly in the fields of sports medicine, human genetics and internal medicine. He also had editorial responsibilities with several German and international medical journals.

Sports medicine was Grebe's major interest and he was president of the Association of German Sports Doctors from 1957 until 1961. He was also a founder member of the German Olympic Society and a member of the scientific committee of the world body for sports and physical education at UNESCO. Grebe was himself a fine athlete, excelling in swimming, athletics and tennis. Indeed, he won a triathlon championship in Frankenberg at the age of 51 years. Grebe also had an interest in amateur boxing and for 30 years from 1957 he was chairman of an international committee which was concerned with the medical aspects of this sport.

Grebe was an official of the Red Cross for most of his professional life and received several merit awards from that body. He was also an active Rotarian, serving as a district governor in 1976 and 1977 and on the World Fellowship Committee of Rotary International from 1980 to 1985.

Grebe married Irmgard Hartmann in 1938 and they had 2 sons, a daughter and 9 grandchildren. In 1995 at the age of 81 years, Grebe was living in retirement in Frankenberg, Germany.

# Nomenclature

G REBE developed an interest in genetic skeletal disorders during the early phases of his career at Berlin and Rostock. He accumulated a vast quantity of material which he further analysed in Frankenberg and while teaching at the University of Marburg, and in 1952 he delineated the condition which now bears his name. His patients, dwarfed sisters aged 7 and 11 years, had unaffected consanguineous parents, and he recognised that the mode of inheritance was probably autosomal recessive.

Grebe employed the designation "achondrogenesis" in the title of his paper and in the text of a review which he wrote in 1955. In 1964 Quelce-Salgado identified 47 affected persons in 5 large consanguineous Brazilian kindreds and thereafter published a series of papers on many aspects of the disorder. By virtue of this geographical clustering, the name "Brazilian achondrogenesis" gained some favour. Thereafter, there was terminological confusion with the forms of lethal short-limbed dwarfism designated as "achondrogenesis IA (Parenti-Fraccaro)" and "achondrogenesis IB (Langer-Saldino)". In this numerical system, the condition described by Grebe was termed "achondrogenesis II (Grebe achondrogenesis, Brazilian achondrogenesis, Grebe chondrodysplasia)".

In the 1983 Paris Nomenclature, the lethal conditions were designated achondrogenesis type I and II respectively. The condition delineated by Grebe is clinically and radiologically distinct from these lethal conditions and the eponymous form "Grebe chondrodysplasia" came into general use. In the 1992 International Classification of Osteochondrodysplasias, Grebe dysplasia was categorised as a form of acromesomelic dysplasia.

The concept of autosomal recessive inheritance was strengthened by reports of affected siblings with consanguineous parents in families in India (Meera Khan and Khan, 1982) and in China (Kumar et al. 1984). In 1986 Curtis drew attention to the fact that obligate heterozygotes may manifest with brachydactyly.

# References


Curtis D (1986) Heterozygote expression in Grebe chondrodysplasia (Letter). Clin Genet 29: 455–456

Grebe H (1952) Achondrogenesis: ein einfach rezessives Erbmerkmal. Folio Hered Path 2: 23–28

Grebe H (1955) *Chondrodysplasie.* Rome: Inst Greg Mendel, pp. 300–303

Kumar D, Curtis D, Blank CE (1984) Grebe chondrodysplasia and brachydactyly in a family. Clin Genet 25: 68–72

Meera Khan P, Khan A (1982) Grebe chondrodysplasia in three generations of an Andhra family in India. In: Papadatos CJ, Bartsocas CS (eds) *Skeletal Dysplasias.* Alan R Liss, New York

Quelce-Salgado A (1964) A new type of dwarfism with various bone aplasias and hypoplasias of the extremities. Acta Genet Statist Med 14: 63–66

# GROENBLAD, Ester Elisabeth

*(1898–1976)*

*Courtesy:* Professor Bjørn Tengroth, Sweden

G ROENBLAD-STRANDBERG disease or pseudoxanthoma elasticum (PXE) is a heterogeneous genetic disorder characterised by yellowish indurated patches in the skin flexures and angioid streaks in the retina. The skin becomes lax in affected areas and ocular involvement may lead to visual impairment.

For many years the conjoined eponym enjoyed favour, especially in the continental literature but the designation "pseudoxanthoma elasticum" (PXE) is now generally preferred. Several hundred affected persons have been reported and it is evident that PXE is clinically and genetically heterogeneous.

# Biography

G ROENBLAD was a Swedish ophthalmologist, active in Stockholm during the twentieth century.

Ester Elisabeth Groenblad was born in Uppsala, Sweden on 1st August 1898. Her father was a successful businessman and her family enjoyed favourable social circumstances. She entered the University of Stockholm in 1916 and qualified in medicine in 1920. Groenblad subsequently trained in ophthalmology at the Seraphine Hospital in that city and after spending a 2-year period visiting overseas ophthalmological centres, she entered private practice. In addition to this commitment, Groenblad was the head of the Ophthalmological Service for Stockholm Schools from 1930 until 1943. She maintained a life-long interest in the social welfare of children and served on the Stockholm School Board for many years.

Groenblad obtained her doctorate in 1933 for a dissertation on angioid streaks in pseudoxanthoma elasticum. This work represented her major medical contribution, although she published other articles on eye disease with skin manifestations. She also collaborated with H Sjögren in investigations which led to the delineation of "Sjögren's disease".

Groenblad had sophisticated cultural interests, and she was friendly with several prominent authors and intellectuals. Her summers were spent at Forusund, a resort outside Stockholm, where she undertook research into local history. These activities culminated in a book and a series of articles on this subject. Groenblad never married; she died in 1976 at the age of 78 years.

# Nomenclature

T HE cutaneous manifestations of the condition now known as pseudoxanthoma elasticum (PXE), or Groenblad-Strandberg disease, were documented by Rigal in 1881 and the autopsy findings were reported by Balzar in 1884. In 1889 Anatole Chauffard gave a detailed description of the skin manifestations of a French soldier, who had suffered from haematemesis during his military service in New Caledonia. The condition was initially regarded as a xanthomatosis, but was differentiated from this group of disorders by Darier in 1896, following histological studies of a skin biopsy specimen from Chauffard's patient. Darier then proposed the designations "pseudoxanthoma elasticum" and "elastorrhexis" on a basis of these histopathological observations. The patient went on to develop visual problems; Hallopeau and Laffitte (1903) documented the retinal changes and made a tentative suggestion that the cutaneous and ocular features might be manifestations of the same basic defect.

The syndromic association of angioid streaks and the skin changes of PXE was firmly established by Groenblad in 1929, in collaboration with her dermatological colleague, James Strandberg (*see* p. 161). Further details were published in 1932 and in 1958, almost 30 years after her original article, Groenblad published colour photographs of the retinal appearances in the late stages of the disorder.

# References

Balzar F (1884) Récherches sur les charactères anatomiques due xanthélasma. Arch Physiol 4 (series 3): 65–80

Chauffard A (1889) Xanthélasma disséminé et symétrique ans insuffisance hépatique. Bull Soc Méd Hop Paris 6 (series 3): 412–419

Groenblad E (1929). Angioid streaks – pseudoxanthoma elasticum: vorläufige Mitteilung. Acta Ophthalmol 7: 329–336

Groenblad E (1932) Angioid streaks – pseudoxanthoma elasticum. Der Zusammenhang Zwischen Diesen Gleichzeitig Auftretenden Augen – und Hautveränderungen. Acta Ophthalmologica Supp I, Vol 10

Groenblad E (1958) Colour photography of angioid streaks in the late stages. Acta Ophthalmologica 36: 472–474

Hallopeau H, Laffitte P (1903) Nouvelle note sur un cas de pseudoxanthome élastique. Ann Derm Syph Paris 4: 595

Rigal D (1881) Observation pour servir à l'histoire de la chéloide diffuse xanthelmique. Ann Derm Syph 2: 491–501

Strandberg J (1929) Pseudoxanthoma elasticum. Z Haut Geschlechtske 31: 689–693

# GUNN, Robert Marcus

## *(1850–1909)*

G UNN phenomenon or familial jaw winking syndrome comprises unilateral ptosis with involuntary elevation of the eyelid on opening the mouth.

# Biography

G UNN was a Scottish ophthalmologist. He became senior surgeon at the Royal London Ophthalmic Hospital, Moorfields.

Robert Marcus Gunn was born in 1850 at Dunnet, Sutherland, in the far north-west of Scotland, where his father was a farmer. He attended the University of St Andrews and qualified in medicine, with distinction, in 1873. By that time he had developed an interest in ophthalmology and he gained postgraduate experience at the Moorfields Eye Hospital, London and under Jaeger at Vienna. He also undertook a locum tenens post at the lunatic asylum in Perth, where he surveyed the retinal appearances of the patients. In 1876 Gunn was appointed as house surgeon at Moorfields and during his tenure of this post he gained a reputation for efficiency, diligence and innovation.

A long-standing interest in the comparative anatomy of the eye led to a visit to Australia in 1879 in order to collect eye specimens from the indigenous species. On his return Gunn was involved in the study of material brought to England by Charles Darwin's famous "Challenger" biological expedition.

After obtaining the fellowship of the Royal College of Surgeons in 1882, Gunn rose through the professional ranks at Moorfields, being appointed as surgeon in 1888. He also received appointments at the Hospital for Sick Children (Great Ormond Street), the National Hospital for the Paralysed and Epileptic, Queen Square, and at University College Hospital. Gunn became vice-president of the British Medical Association's section of ophthalmology, representing this group at meetings in Edinburgh, Toronto and the USA.

Gunn had many personal qualities, including a high intellect, clarity of thought, a good memory, intellectual honesty and balanced judgement. He had a modest disposition which concealed strength of character and an element of intolerance. He maintained exemplary standards in his personal and professional life and he was always courteous and cordial. Gunn's professional reputation was based on his abilities at ophthalmoscopy and, to a lesser extent, on his surgical skills and the quality of his teaching. He was also a typical Victorian polymath, with interests in botany, marine biology, zoology and geology; his holidays were spent seeking fossils and prior to his death, he donated his large collection to the British Museum.

Gunn's health began to fail in 1907 and he died in London in 1909, at the age of 59 years, following several bouts of influenza.

# Nomenclature

I N 1883 Gunn published an article entitled *Congenital ptosis with peculiar associated movements of the affected lid*. The disorder is evidently not uncommon and by 1895, Sinclair was able to review the manifestations in 33 affected persons. The unusual stigmata precluded any simple descriptive designation but this problem was resolved in 1925 when Villard, writing in the French literature, introduced the eponymous term "Marcus Gunn" phenomenon. This title is still retained, with the occasional erroneous insertion of a hyphen. Gunn's first forename was Robert, and it seems that he followed the contemporary style of using his second forename as an addition to his surname (as, for instance, did his colleague, Edward Treacher Collins). This format had developed by necessity in earlier times, when the parents of an heiress sometimes insisted on the hyphenated retention of their family name and it evolved into a mark of social distinction, which was used by the upper social classes in the Victorian era. In view of Gunn's lack of ostentation it is surprising that he chose to follow this convention and it is possible that it was thrust upon him by colleagues who subsequently quoted his name in their articles.

Despite the interest that the unusual manifestations of the Marcus Gunn phenomenon has generated, the pathogenesis is still uncertain. It is possible that there is an abnormal anatomical connection involving the motor branch of the trigeminal nerve supply to the levator palpebrae superioris muscle. A familial propensity has been documented on several occasions and pedigree data are suggestive of autosomal dominant inheritance, with incomplete penetrance.

Many articles concerning the condition have appeared in the literature, and alternative designations include "maxillopalpebral synkinesis", "hereditary palpebromaxillary synergy", "synketic ptosis", "pterygoid-levator synkinesis" and the "jaw-winking syndrome". The latter has the virtue of simplicity but the eponymous format "Marcus Gunn" is firmly established, perhaps because it trips easily off the tongue.

An "inverted Marcus Gunn syndrome" or "reverse jaw-winking syndrome" was documented in 1918 in the Spanish literature by an ophthalmologist, Manuel Marin Amat. In this condition the eye closes when the mouth is opened widely. The mode of genetic transmission, if any, is uncertain.

# References


Falls HF, Kruse WT, Cotterman CW (1949) Three cases of Marcus Gunn phenomenon in 2 generations. Am J Ophthalmol 32: 53–59

Gunn M (1883) Congenital ptosis with peculiar associated movements of the affected lid. Trans Ophthalmol Soc UK 3: 283–287

Marin Amat M (1918) Contribucion al estudio de la curabilidad de las paralisis oculares de origen traumatico-substitucion funcional del VII por el V par craneal. Arch Oft Hisp Amer 18: 70–99

Obituary (1909) Br Med J 2: 1719–1721

Sinclair WW (1895) Abnormal associated movements of the eyelids. Ophthalmol Rev 14: 307–311

Villard H (1925) Le phénomene de Marcus Gunn. Bull Soc Franç Ophthalmol 38: 725–735

# HALL, Judith G
*(b. 1939)*

H ALL type of pseudoachondroplasia is characterised by severe dwarfism, limb shortening and variable spinal malalignment.

# Biography

H ALL is a leading North American dysmorphologist with an international reputation in her field.

Judith Hall was born on 3 July 1939, in Boston, USA where her father was a clergyman. She obtained a BA degree at Wellesly College, Massachusetts, in 1961 and proceeded to the University of Washington, Seattle, where she qualified in medicine in 1966. During her medical studies Hall spent a year in the University genetics department and received the master of science degree for a thesis entitled *In vitro fetal hemoglobin synthesis.*

Hall began her medical career at the Johns Hopkins Hospital, Baltimore, as a Fellow in medical genetics with Professor V. A. McKusick, and subsequently as a resident in paediatrics. In 1972, after completing her specialist training, Hall returned to the University of Washington School of Medicine, where she had a fruitful collaboration with the late Professor David Smith. Her academic achievements received recognition in 1980 when she was appointed as professor of medicine and paediatrics in the division of medical genetics at the University of Washington. In 1981 Hall moved to the University of British Columbia, Canada, to the post of professor of medical genetics and in 1990 she became chairman of the department of paediatrics.

After her early work on the delineation of various forms of dwarfism, Hall became interested in stiff joint conditions. Her investigations resulted in the description of several distinct entities within this broad category and the clarification of nosological concepts concerning arthrogryposes. Dysmorphology became the focus of her academic attention and she developed the concept of long-term follow-up and the pictorial documentation of the natural history and age relationships of the manifestations of dysmorphism syndromes. In this way new insights have been obtained into the pathogenesis of these disorders.

Hall is the author of more than 200 medical articles and chapters and she received recognition for her achievements in the form of visiting professorships at universities throughout the world. She has also received awards as an outstanding teacher and she has been an invited speaker at many congresses. Besides these activities, she has been a member of the controlling bodies of several professional and paramedical organisations and has held editorial responsibilities with a number of prominent medical journals.

Hall has a highly developed sense of humanity, as evidenced by her statement "to me, high achievement is not the number of publications but being a successful female in a world of professional men. And by that I mean caring more about peacemaking and nurturing the individual and the environment than success, winning, owning or directing".

# Nomenclature

D URING her fellowship in the Moore Clinic, Johns Hopkins Hospital, Baltimore, Hall became interested in the heritable disorders of connective tissue. At the time she was involved in the delineation of dwarfing skeletal dysplasias, including the autosomal recessive form of pseudoachondroplasia, which now bears her name.

Until the middle years of the twentieth century, dwarfism was broadly divided into the short limb and short trunked forms, with the eponym "Morquio syndrome" being loosely applied to any dwarfing disorder with significant spinal involvement. At the end of the 1950s, with the advent of modern genetic concepts, it became evident that there was great heterogeneity amongst the skeletal dysplasias and numerous specific syndromes were delineated at this time. Amongst these was a condition termed "pseudo-achondroplastic spondyloepiphyseal dysplasia", which was differentiated from classical achondroplasia by Maroteaux and Lamy (1959). In turn this condition was differentiated from spondyloepiphyseal dysplasia and termed "pseudo-achondroplasia".

Organisations of persons with small stature, notably the "Little People of America". provided researchers with the opportunity to examine large numbers of affected persons. In this way it became apparent that pseudoachondroplasia was a comparatively common dwarfing skeletal dysplasia. Hall became involved with the "Little People of America" during her research fellowship at the Johns Hopkins Hospital and she investigated the genetic basis of this disorder. She was subsequently accorded the unique honour of life membership of this organisation. She recognised phenotypic variability at the clinical and radiological level, together with genealogical evidence for AD and AR forms, and postulated that the condition was heterogeneous. Four separate entities were delineated, 2 AD and 2 AR, with mild and severe forms of each. Hall's name was attached to the severe AR form of the disorder.

Recognition of variation of phenotypic manifestations in the same family have clouded the issue of autonomy of sub-types of pseudoachondroplasia and in the 1983 International Nomenclature of Constitutional Disease of Bone (Paris Nomenclature) AD and AR forms of pseudoachondroplasia were listed without further subdivision and without eponyms. The trend towards lumping rather than splitting continued, and in the 1992 International Classification of Osteochondrodysplasias, pseudoachondroplasia was listed as a single AD entity. In 1994 the demonstration by a consistent linkage to a locus on chromosome 19 in several affected kindreds further strengthened the concept of genetic homogeneity.

In addition to pseudoachondroplasia, Hall's name is linked eponymously with that of Pallister in the title of a rare congenital syndrome comprising hamartoblastoma of the hypothalamus, dysplasia of the lungs and kidneys, cardiac defects and growth retardation (*see* Pallister, p. 217).

# References


Hall JG, Dorst JP (1969) Four types of pseudoachondroplastic spondyloepiphyseal dysplasia (SED). Birth Defects Orig Art Ser V: 242–257

Hall JG (1975) Psuedoachondroplasia. Birth Defects Orig Art Ser XI: 187–202

Hall JG, Grossman M (1969) Case report F2 – Pseudoachondroplastic SED. Birth Defects Orig Art Ser V: 207–210

Hall JG (1974) Pseudoachondroplasia (pseudoachondroplastic spondyloepiphyseal dysplasia). Am J Dis Child 128: 833–834

Hall JG, Pallister PD, Clarren SK, Beckwith JB, Wiglesworth FW, Fraser FC, Cho S, Benke PJ, Reed SD (1980) Congenital hypothalamic hamartoblastoma, hypopituitarism, imperforate anus and postaxial polydactyly – a new syndrome? Am J Med Genet 7: 47–74

Hecht JT, Francomano CA, Briggs MD, Deere M, Connor B, Horton WA, Warman M, Cohn DH, Blanton SH (1993) Linkage of typical pseudoachondroplasia to chromosome 19. Genomics 18: 661–666

Maroteaux P, Lamy M (1959) Les formes pseudoachondroplastiques des dysplasies spondyloepiphysaires. Press Méd 67: 383–386

# HANHART, Ernst
## *(1891–1973)*

*Courtesy*: Professor D. Klein, Switzerland

H ANHART dwarfism is an autosomal recessive form of panhypopituitarism which is characterised by proportionate short stature and hypogonadism.

# Biography

H ANHART was a Swiss physician during the first half of the twentieth century and a founder of medical genetics in his country.

Ernst Hanhart was born in Switzerland on 14 March 1891 and qualified in medicine at the University of Zurich in 1916. He initially became a country practitioner in the Canton of Argovie but, seeking intellectual stimulation, moved to Zurich in 1921 as assistant at the Policlinic under Professors O. Naegeli and W. Löffler.

Hanhart developed a profound interest in human genetics and rapidly became a specialist in hereditary disorders. He recognised the importance of investigating large affected families and he was one of the first to study endogamous genetic isolates. In this context he drew attention to the fact that the offspring of incestuous relationships were of special significance in the study of the effects of consanguinity. Hanhart made early contributions concerning the genetics of many disorders, including Friedreich ataxia, familial deafness, mental retardation, mongolism, schizophrenia, diabetes and pituitary dwarfism (vide infra).

The results of Hanhart's investigations in Swiss isolates were published in an article dedicated to the memory of Gregor Mendel on the 90th anniversary of the discovery of the laws of inheritance. In this work he presented details of 31 disorders which he had studied in the Swiss population. Hanhart eventually wrote more than 150 medical articles, the majority of which concerned genetic conditions, and he was also the author of several chapters on human mutations and hereditary metabolic disorders.

Hanhart's contributions were recognised with his appointment as titular professor at the University of Zurich in 1942. In this period he was involved in the formation of the Swiss Society of Genetics, which had the aim of encouraging genetic research for the sake of pure science, and not for political or racial considerations. This Society was formed as a response to developments in the science of genetics in Nazi Germany.

By the end of World War II, Hanhart had achieved an international reputation. He was a member of many scientific organisations and received frequent invitations as a congress speaker. He was noted for his elegant and persuasive style of delivery, and for the scientific accuracy of his presentations.

Hanhart's career was compromised in 1954 when a criminal attack resulted in severe physical handicap. He was obliged to take premature retirement from his University and he moved to Ascona, where he continued with his scientific writings. In 1971, on the occasion of his 80th birthday, Hanhart received an eulogy from Professor Prader, on behalf of the medical faculty of Zurich University. He died on 5 September 1973, after a long and painful illness.

# Nomenclature

H ANHART realised the importance of endogamy in the pathogenesis of genetic disease and the study of isolated populations in high Alpine valleys was a major facet of his life's work.

In two villages in western Switzerland (Oberegg and Appenzell with Samnaun in Grisons) Hanhart recognised a form of proportionate dwarfism associated with hypogonadism and realised that the high rate of consanguinity and the occurrence of affected siblings were indicative of autosomal recessive inheritance. His observations were published in 1925 in the German language, under the title *Heredito-degenerative dwarfism with dystrophia adiposogenitalis.* Hanhart's findings also formed the subject of a thesis which was submitted in the same year. He retained a long-standing interest in this condition and he subsequently studied the condition on the Yugoslavian island of Krk (Veglia) in the Adriatic.

As reports of pituitary dwarfism accumulated, it became evident that the disorder was very heterogeneous. The genetic forms, which are in the minority, were given numerical and descriptive designations; the entity which formerly bore Hanhart's name is now conventionally entitled "pituitary dwarfism type III (panhypopituitarism)".

In addition to pituitary dwarfism, Hanhart's eponym is attached to an autosomal recessive disorder in which amelia of the upper, or all four, extremities is associated with micrognathia. It is also used for an autosomal dominant form of palmar-plantar hyperkeratosis and for autosomal recessive familial spastic paraplegia with mental retardation.

# References

Bersu ET, Petersen JC, Charboneau WJ, Opitz JM (1976) Studies of malformation syndromes of man XXXX1A: Anatomical studies in the Hanhart syndrome – a pathogenetic hypothesis. Europ J Pediat 122: 1–17

Hanhart E (1925) Ueber heredo-degenerativen Zwergwuchs mit Dystrophia adiposo-genitalis. Arch Klaus Stift, Zurich 1: 181–257

Hanhart E (1947) Neue Sonderformen von Keratosis palmo-plantaris, u.a. eine regelmassig-dominante mit systematisierten Lipomen, ferner 2 einfach-rezessive mit Schwachsinn und z.T. mit Hornhautveränderungen des Auges. Dermatologica, Basel 94: 286–308

Hanhart E (1950) Uber die Kombination von Peromelie mit Mikrognathie, ein neues Syndrom beim Menschen, entsprechend der Akroteriasis congenita von Wriedt und Mohr beim Rinde. Arch Julius Klaus Stift, Zurich 25: 531–544

Hanhart E (1953) Die Rolle der Erbfaktoren bei den Stoerungen des Wachstums. Schweiz Med Wschr 83: 198–203

Obituary (1973) Klein D. Medicine and Hygiene 1071: 1417

Prader A (1971) Eulogy. Prof Ernst Hanhart 80th year. Schweiz Med Wschr 101/11: 402

# HOLLISTER, David William

## *(1941–1991)*

H OLLISTER syndrome, or "long thumb brachy-dactyly" is an autosomal dominant disorder comprising digital shortening, short clavicles, thoracic ₃try and cardiac anomalies.

# Biography

H OLLISTER was an outstanding medical scientist in the USA. He made important contributions to the understanding of the heritable disorders of connective tissue, notably the Marfan syndrome.

David Hollister, the son of William Hollister, a psychiatrist, was born in Omaha, Nebraska on 18 September 1941. He studied at Oberlin College and qualified in medicine at Duke University. After internship at Duke University Hospital, Hollister gained experience in laboratory work at the National Institute of Arthritis and Metabolic Disorders. He then undertook residency training in internal medicine at the Barnes Hospital, St Louis. By this time he had developed a keen interest in heritable disorders of the skeleton and connective tissue and he became the first fellow in medical genetics at the University of California, Los Angeles (UCLA). Hollister's outstanding potential was recognised and in 1971 he was granted faculty status and headship of a connective tissue laboratory at UCLA.

A decade later, Hollister took a year's sabbatical at the Max Planck Institute, Munich, Germany and in 1983 he moved with members of his research team to Portland, Oregon, where he set up a multidisciplinary research department at the Shriners Hospital. In his final move Hollister returned to his roots in Omaha, where he was appointed as Head of the Connective Tissue Diseases Laboratory at the Munroe Center for Human Genetics with the status of professor of paediatrics, pathology, microbiology and medicine.

Hollister's main research directions were in the production of monoclonal antibodies to connective tissue proteins, and the identification of collagens, with the ultimate aim of the determination of the basic defects in genetic disorders of skin, bone and joints. Hollister retained his clinical skills throughout his career and he delineated several new disorders, including a condition documented in collaboration with his father.

In the years which preceded Hollister's move to Omaha, the Marfan syndrome was the focus of considerable interest in the field of medical genetics. It had been confidently predicted that this disorder would probably be one of the first in which the basic defect would be recognised; as things turned out, it was amongst the last to be elucidated. Hollister and his team played a key role in this process by virtue of their discovery of abnormalities of the connective tissue component, fibrillin, in affected individuals. This work led to the chromosomal localisation of the fibrillin gene, the establishment of linkage to the Marfan syndrome and the detection of fibrillin gene mutations in persons with the disorder. Hollister's contribution was recognised by the posthumous bestowal of the Antoine Marfan Award and, as a further gesture, the National Marfan Foundation established the David Hollister research fellowship. The laboratory at the University of Nebraska has been named in his honour, and a plaque commemorating his memory hangs on the laboratory wall.

David Hollister was an extremely genial and likeable person. He had a reserved demeanour, a quiet smile, and he radiated warmth and goodwill. In the highly competitive world of medical research, interpersonal relationships are often strained, but Hollister was universally well-liked. He was also held in high regard for his medical and scientific abilities; his skills ranged through clinical genetics, radiology, histology, biochemistry, immunology and molecular biology. Not surprisingly, he had an international reputation in his field.

Hollister developed carcinoma of the pancreas in 1990 and his highly productive career came to an untimely end with his death on 3 February 1991 at the age of 50 years.

# Nomenclature

I N 1981, while undertaking research into genetic connective tissue disorders at Harbour-UCLA medical centre, David Hollister collaborated with his father, William G. Hollister, a psychiatrist at the University of North Carolina, Chapel Hill, in the delineation of the condition which bears their conjoined eponym.

The family whom they described comprised a brother, sister and father with brachydactyly, mild skeletal abnormalities, articular rigidity and cardiac conduction defects. The Hollisters noted that in contrast to the shortening of the fingers, the thumbs were comparatively long. The affected father and his own affected mother had been investigated in 1939 or thereabouts by David Whitney of the Department of Zoology, University of Nebraska. Whitney had suggested that the condition was a genetic trait, but apparently his findings were never published.

In their article Hollister and Hollister termed the condition "the long thumb brachydactyly syndrome" and discussed the differential diagnosis from other brachydactyly and clavicular dysplasia disorders. No previous reports could be found in the literature and it was evident that the condition was a "private syndrome". The 10th edition of McKusick's *Mendelian Inheritance in Man* lists the disorder as "brachydactyly, long-thumb type", without quoting additional cases, and without the asterisk of proven syndromic identity.

In addition to the long thumb brachydactyly, Hollister's name is also associated with the lacrimo-auriculo-dento-digital syndrome. The first description of the disorder in 1967 is possibly attributable to Walter J. Levy, an ophthalmologist of Johannesburg, South Africa, who published an account of a single patient with a disorder which he termed "mesoectodermal dysplasia". In 1973 Hollister and several colleagues reported a Mexican family in which a father and 5 of his 8 children had abnormalities of the external ears, teeth, lacrimal ducts and digits. These authors proposed the present title "lacrimo-auriculo-dento-digital syndrome" but the conjoined eponym Levy-Hollister syndrome is now sometimes used.

# References

Hollister DW, Klein SH, Dejager HJ, Lachman RS, Rimoin DL (1973) The lacrimo-auriculo-dento-digital syndrome. J Pediatr 83: 438–444
Hollister DW, Hollister WG (1981) The "long-thumb" brachydactyly syndrome. Am J Med Genet 8: 5–16
Levy WJ (1967) Mesoectodermal dysplasia: a new combination of anomalies. Am J Ophthalmol 63: 978–982
Obituary. Am J Med Genet 43: 511512

# HOLMES, Sir Gordon
## *(1876–1965)*

*From*: Parsons-Smith G.B. (1982) In: Rose F.C., Bynum W.F. (eds) Historical Aspects of the Neurosciences

*Courtesy*: Lippincott-Raven Publishers, Philadelphia

DIE-HOLMES syndrome (*see* Adie, p. 5).

# Biography

HOLMES was a leading London neurologist during the first half of the twentieth century.

Gordon Morgan Holmes was born on 22 February 1876 in Dublin, Ireland. His father, Gordon Holmes, was a successful farmer at Dellin House, Castlebellingham, County Louth, Ireland, about 40 miles north of Dublin. The early death of his mother, Kathleen, née Morgan, and his father's remarriage, deeply affected Holmes, and although he had 3 brothers and 3 sisters, he was a solitary child. Despite a transient dyslexia, Holmes was a brilliant scholar and after completing his education as a boarder at Dundalk academy, Holmes entered Trinity College, Dublin and graduated in medicine in 1897 at the age of 21 years.

In his youth Holmes was a tall, handsome athletic man, with dark, curly hair and a taste for adventure. Soon after qualification he worked his passage to New Zealand, serving as ship's surgeon. He then took up a scholarship in Germany, where he spent 2 years studying neuroanatomy at the University of Frankfurt, under Professor Ludwig Edinger. Holmes decided to specialise in clinical neurology, and after his return to England he obtained a post at the National Hospital, Queen Square, London, as house physician to Hughlings Jackson, the doyen of British neurologists.

Holmes made a good impression at Queen Square and in 1906 he was appointed as director of clinical research. He developed an interest in the anatomical connections of the tracts in the spinal cord, and he published extensively in this field. Holmes still retained his urge for adventure and he sought a place on Captain Scott's ill-fated exploration expedition to the South Pole. A ruptured Achilles tendon necessitated the abandonment of this plan and Holmes profited from his convalescence by obtaining a higher medical degree. An unexpected vacancy on the staff of the National Hospital arose in 1910 when a senior colleague died suddenly, and with the kudos of his recent degree, Holmes was elected to this post. Thereafter his life revolved around his clinical and teaching activities in this hospital, which were unpaid, and a successful private practice in Harley Street. As an antidote to his hectic medical life Holmes enjoyed long summer holidays in his old home in Ireland where he coached the village tug-of-war team. He also derived pleasure from taking his resident medical staff for an annual 3-day row along the river Thames from Oxford to London. His residents' comments on this arrangement have not been recorded!

When World War I broke out in 1914 Holmes was appointed as consultant neurologist to the British Army in France and together with his neurosurgical colleague, Percy Sargent, he set up a field hospital for the treatment of wounds to the head. Many hundreds of wounded soldiers were successfully treated and their injuries provided Holmes with a unique opportunity for the investigation of the effects of lesions in specific regions of the brain on balance, vision and bladder function. While in France, Holmes met his future wife, Dr Rosalie Jobson, an Oxford graduate and an international sportswoman, to whom he subsequently proposed marriage while rowing on the Thames. Holmes had a happy married life and he was blessed with 3 daughters and a gracious home. He enjoyed a biweekly game of golf with his wife, which he usually lost, and a regular evening reading session, which he devoted to the works of Shakespeare, Thackeray and Tennyson.

In the period between the wars, Holmes had concurrent appointments at Queen Square, Moorfields Eye Hospital and the Charing Cross Hospital. He was an exceptional teacher of clinical neurology and his weekly case discussions at Queen Square attracted numerous postgraduates. Holmes had a powerful physique, beetling brows and a piercing gaze, his inner warmth was masked by an abrupt manner, and he was held in awe by his students and junior staff but despite these intimidating attributes, he was esteemed and widely respected. His biographer, Parsons-Smith, stated "His ward round took place behind locked doors. His nursing staff, who thought he was wonderful, saw to it that the patients lay still; they were not allowed to read a paper as the rustle might prove distracting. Only a jug of water with a glass was permitted on the locker top and, though in certain cases a vomit bowl might be provided, it was not expected to be touched; the use of a bedpan was not even to be contemplated." Parsons-Smith went on to say of his teaching rounds, "He might grasp a worried student by the lapels of his coat and gently rock him backward and forward in rhythm with his instructions."

Holmes received the CMG and CBE for his activities in World War I. His academic achievements led to his election to the Fellowship of the Royal Society and the award of several honorary degrees and his overall contributions to neurology were acknowledged with a knighthood.

During World War II Holmes joined the Emergency Medical Service as adviser in neurology and continued his teaching at the Charing Cross Hospital, which had been evacuated to Ashridge. He eventually settled in Farnham, Surrey, where he enjoyed a long retirement. Holmes died in his sleep on 29 December 1965, at the age of 90 years.

# Nomenclature

AT the end of World War I Holmes returned to the staff of the Charing Cross Hospital and was joined by William Adie, a young Australian who had served in France in the head injuries unit. Adie had considerable intellectual gifts and he rose through the hospital ranks, eventually receiving consultant appointments at Charing Cross Hospital and Queen Square.

Holmes and Adie became close friends and they shared academic interests in neurology and neuroanatomy. In 1931 they published separate papers on the condition which now bears their conjoined eponym. It seems probable that they had entered into extensive discussion of the disorder which they had documented and it is fitting that their names are linked for posterity (*see* Adie, p. 5).

# References

Adie WJ (1931) Pseudo-Argyll Robertson pupils with absent tendon reflexes. A benign disorder simulating tabes dorsalis. Br Med J 1: 928–930

Holmes G (1931) Partial iridoplegia associated with symptoms of other disease of the nervous system. Tr Ophth Soc UK 51: 209–228

Obituary (1966) Br Med J 1: 111–112

Obituary (1967) J Neurol Sci 5: 185–192

Parsons-Smith BG (1982) Sir Gordon Holmes. In: Rose FC, Bynum WF (eds) *Historical Aspects of the Neurosciences*. Raven Press, New York, pp. 357–370

# HOOFT, Carlos M.

*(1910–1980)*

*Courtesy*: Professor Jules Leroy, Belgium

H OOFT syndrome, or familial hypolipidaemia, comprises postnatal growth retardation, an erythematous eruption on the face and limbs, abnormalities of the nails, hair and teeth and inconsistent retinal changes. Affected individuals have a characteristic biochemical profile. Inheritance is autosomal recessive.

## Biography

H OOFT was professor of paediatrics and Dean of the medical faculty of the University of Ghent, Belgium.

Carlos M. Hooft was born in Belgium on 15 March 1910 and qualified in medicine at Ghent University Medical School in 1935. He was a talented student and received an inter-university fellowship award for a research paper entitled *Abnormal functional changes in connective tissue*. Hooft trained in paediatrics in Leiden and Ghent and at the Hôpital pour Enfants Malades, Paris. In 1939 he successfully defended his doctoral thesis on the topic "Euglobulinaemia and Pseudoglobulinaemia" and he then became Chairman of the paediatric department of the University of Ghent. He was elevated to full professorship of paediatrics in 1948, a status which he retained until his sudden death in 1980. He was succeeded as chairman by Jules Leroy (*see* p. 215), who still occupied that post in 1996.

Hooft was active in his medical faculty and served as Dean during the period 1961–1970. He was also a member of the Academies of Medicine and Science of Belgium and the Netherlands, and the Paediatric Societies of Belgium, the Netherlands and France. Hooft became a titular member of the Royal Academy in 1950 and served as President in 1973–4.

Many research contributions were made by Hooft in the broad field of paediatrics. In particular he undertook extensive and original work concerning lipoidnephrosis, plasma protein malnutrition, tuberculous meningitis, poliomyelitis, congenital malformation syndromes and inborn errors of metabolism. He was the first to recognise hypothyroidism in cystinotic children.

Hooft had exceptional clinical skills and he was a noted teacher. During his long career he succeeded in passing on his clinical experience and scientific eagerness to several generations of medical students and paediatricians.

## Nomenclature

H OOFT had a special research interest in the inborn errors of metabolism and in 1962, together with his colleagues at the University of Ghent, he published an account of 2 sisters with a familial hypolipidaemia growth retardation syndrome. One of these girls also had tapetoretinal degeneration, and both had mental retardation. They also had erythematous skin lesions and abnormalities of the nails, hair and teeth. Biochemically, serum lipid levels were low, metabolism of aminoacids was faulty and renal tubular reabsorption of phosphate was increased. In the following year Francois and De Blond commented upon the familial nature of the disorder and emphasised the association between the retinal changes and reduced levels of lipids in the serum.

Hooft disease, or familial hypolipidaemia, as the disorder is now known, is similar in many ways to abeta-lipoproteinaemia, which is well-defined and much more common. The disorders differ by virtue of the lack of malabsorption and acanthocytosis in Hooft disease, and the biochemical profiles are also distinct.

## References


Francois J, De Blond R (1963) Dégenerescence tapeto-retinienne associée à un syndrome hypolipidemique. Acta Genet Med Gemellol 12: 145–157

Hooft C, De Laey P, Herpol J, DeLoore F, Verbeeck J (1962) Familial hypolipidaemia and retarded development without steatorrhoea. Another inborn error of metabolism? Helvet Paediatr Acta 17: 1–23

Obituary (1980) "Professor Carlos Hooft" Acta Paediatr Belg 33(2): 75–76

Salt HB, Wolff OH, Lloyd JK, Fosbrooke AS, Cameron AH, Hubble DV (1960) On having no beta-lipoprotein: a syndrome comprising abeta-proteinaemia, acanthocytosis and steatorrhoea. Lancet II: 325–329

# HORNER, Johann Friedrich

*(1834–1886)*

H ORNER syndrome comprises unilateral ptosis and abnormal pupillary reactions, due to dysfunction of the cervical sympathetic nerves. The condition is usually non-genetic, but a rare congenital autosomal dominant form has been documented.

# Biography

H ORNER was Professor of Ophthalmology at the University of Zurich in the second half of the nineteenth century.

Johann Friedrich Horner was born in Zurich, Switzerland on 27 March 1831. His father was a physician and his mother was a talented linguist, from whom he learnt several languages. Horner received his schooling in Zurich, and following a period of compulsory military service, he enrolled at the University Medical School in 1849. He obtained his doctorate in 1854 with a highly praised thesis on the subject of spinal curvature. After obtaining experience in Munich and Vienna, Horner became friendly with von Graefe (*see* p. 179), the pre-eminent ophthalmologist of his era, who influenced his choice of career. Horner then moved to Paris, where he was associated with Desmarres and investigated the role of retinoscopy in the diagnosis of systemic diseases. A call to the Chair of Ophthalmology at the University of Zurich followed and he began an active research career which resulted in the publication of many articles in his field.

In addition to the clinical and research aspects of ophthalmology, Horner recognised the importance of the social implications; in particular, in 1865 he was involved in the enforcement of legislation for the prevention of ophthalmia neonatorum.

Horner died in Zurich on 20 December 1886, at the age of 53 years, while still serving as Professor of Ophthalmology.

# Nomenclature

C HANGES in pupillary reactions, ptosis and elevated temperature of the ipsilateral ear, which follow damage to the cervical sympathetic nerves, are collectively known as the Horner syndrome or Horner's oculo-pupillary syndrome. These abnormalities were, in fact, first demonstrated in 1727 by Francois Pourfour du Petit, by means of experiments involving the transection of the inter-costal and sympathetic nerves in dogs. In 1838, while employed as a house surgeon at the Staffordshire Royal Infirmary, England, Edward Sellick documented similar manifestations in a patient with a tumour in the cervical region. Thereafter, Claude Bernard of France (1862) made similar observations on the effects of stimulation or damage to the cervical sympathetic chain. Silas Weir Mitchell of the USA had been involved in Bernard's investigations and in 1864 he published a review of his findings in his monograph on the neurological sequelae of gunshot wounds and other forms of trauma. Despite these prior publications the definitive description is usually credited to Horner, who in 1869 wrote an account of a woman aged 40 years who developed the classical manifestations of the syndrome.

Although the vast majority of persons with the Horner syndrome have an underlying primary cause for their disability, there have been a few reports of a familial form of the condition. In 1958, using the title "Congenital hereditary Horner's syndrome", Durham documented a family in which 5 persons in 2 generations of a family were affected, it seems likely that this rare genetic form of the disorder is inherited as an autosomal dominant trait.

# References


Bernard C (1862) Des phénomènes oculopupillaires produits par la section du nerf sympathique cervical; il sont indépendants des phénomènes vasculaires colorifique de la tête. C R Acad Sci Paris, 55: 381–388

Birch C (1979) *Names we Remember.* Ravenswood, Beckenham, Kent, England, pp. 70–72

Duke-Elder S (1962–1967) *System of Ophthalmology* Vol XII: 631–632 (1971) Neuro-ophthalmology. Henry Kimpton, London

Durham DG (1958) Congenital hereditary Horner's syndrome. Arch Ophthalmol 60: 939–940

Horner JF (1869) Über eine Form von Ptosis. Klin Mbl Augenheilk 7: 193–198

Mitchell SW, Morehouse GR, Keen WW (1864) Gunshot Wounds and Other Injuries of Nerves. Lippincott, Philadelphia

Pourfoir du Petit F (1727) Mémoire dans lequel il est démontré ques les nerfs intercostaux fournissent des rameaux que portent des espirits dans les yeux. Hist Acad Roy Sci (Paris) (Mém) 1–19

# HUËT, Gauthier Jean

*(1879–1970)*

From: Nederlands Tijdschrift voor Geneeskunde (1954) 98: 3268

Courtesy: Nederlands Tijdschrift voor Geneeskunde,
Amsterdam

PELGER-HUËT Familial white cell anomaly (*see* Pelger, p. 131).

# Biography

H**UËT was a Dutch paediatrician and superintendent of a sanatorium for children.**

Gauthier Jean Huët was born on 23 November 1879 at Amersfoort, Holland. He entered the University of Leiden in 1894, qualified in medicine in 1904 and obtained his doctorate cum laude in 1907 for a thesis on sepsis.

Huët entered private paediatric practice in The Hague and he subsequently became medical superintendent of a children's sanatorium at Hoog-Blaricum, a post which he occupied for almost 50 years. Huët had a special interest in tuberculosis and, although he lacked close academic links, he published many articles on this subject. Huët was also a consultant at the Heideheuval Institute for asthmatic children; the plight of children with severe asthma aroused Huët's sympathy and he devoted considerable effort to the investigation, treatment and psychosocial management of this condition. Amongst his little patients with chest disease was a child with pancreatic dysfunction; Huët's case description proved to be seminal in the development of concepts concerning the genetic condition which is now known as cystic fibrosis.

Huët had a warm personality and a sparkling sense of humour. He was an excellent clinician and radiated warmth to the children who were his patients. His colleagues were impressed by the critical spirit in which he approached scientific problems, and by the breadth of his cultural background. Despite the fact that he did not have an academic appointment, Huët was elected as president of the Dutch Paediatric Association. He was also a popular guest at meetings of the British Paediatric Association.

Huët retired from the Hoog-Blaricum Sanatorium in 1954 when he was 75 years of age. He retained his intellectual faculties and for many years he continued to take an interest in the question of a possible genetic component in asthma. In 1964, in order to mark Huët's sixtieth year in medicine, his colleague S van Creveld wrote: "At his fiftieth jubilee this respected and well-loved man was praised for his extraordinary services. Since then the activities of this youthful old man have not suffered. His writing continued and a conversation with him remained a privilege because of his sparkling and refined spirit. May he still enjoy many healthy and happy years with his spouse."

Huët died in Bussum, Holland, on 13 December 1970, at the age of 91 years.

# Nomenclature

DURING a medical meeting in 1927, Pelger described morphological changes in the nuclei of white cells from 2 children who had died with splenomegaly and tuberculosis. He considered that these leucocytic anomalies were related to the children's condition and his observations were published in the following year. Huët subsequently recognised that the white cell changes were familial and innocuous and published on this point in 1931. Thereafter autosomal dominant transmission of this harmless anomaly was well recognised and the conjoined eponym became firmly established.

The homogeneity of the Pelger-Huët anomaly was questioned in 1974, when Murros and Konttinen documented a kindred in which 4 sisters had familial Mediterranean fever plus the characteristic white cell changes. Several first degree relatives also had leucocytic involvement, and it was suggested that the condition was an autosomal recessive trait, with the full syndrome in homozygotes and partial expression in the white cells of heterozygotes.

In another family 4 brothers with the same mother and 2 different fathers had a Pelger-Huët-like anomaly of their white blood cells, plus abnormalities of the lymphatic system and a propensity to viral and bacterial infections. This condition seems to be an autonomous X-linked trait.

# References

Heyne K (1976) Konstitutionelle familiaere Leukocytopenie mit partieller Pelger-Anomalie und ossaerer Entwicklungsverzoegerung. Europ J Pediat 121: 191–201

Huët GJ (1932) Over een familiaire anomalie der leucocyten. Mschr Kindergeneesk 1: 173–181

Murros J, Konttinen A (1974) Recurrent attacks of abdominal pain and fever with familial segmentation arrest of granulocytes. Blood 43: 871–874

Obituary (1971) In Memoriam: Dr GJ Huët. Maandschr Kindergeneesk 39: 41–42

Pelger K (1928) Demonstratie van een paar zeldzaam vorkomende typen van bloedlichampies en besprecking der patiënten. Ned Tijdschr Geneeskd 72: 1178–1182

Van Creveld (1954) Personalia. Dr GJ Huët 50 Jaar Arts. Ned Tijdschr Geneeskd 98: 3267–3268

Van Creveld S (1964) Dr GJ Huët 60 jaar arts. Ned Tijdschr Geneeskd 108/45: 2191.

# HÜNERMANN, Carl

## *(1900–1943)*

C ONRADI-HÜNERMANN syndrome is a bone dysplasia characterised by radiologically apparent stippling of the epiphyses in infancy, together with shortening and asymmetry of the limbs, depression of the nasal bridge, cataracts and dyskeratotic dermal lesions.

# Biography

H ÜNERMANN was a German paediatrician. His career was prematurely curtailed when he was killed in action in Russia in 1943.

Carl Hünermann was born in Koblenz on 27 October 1900, being the fifth child of Dr Rudolf Hünermann, a general practitioner. He studied medicine in Munster, Munich and Freiburg and qualified in 1924, presenting a thesis on the subject of the therapeutic effects of Vitamin A. Hünermann undertook internships in Berlin, Dusseldorf and Kiel and in 1926 worked in Hanover during a typhus epidemic. He commenced training in paediatrics in 1928 at the University Children's Hospital, Köln, under Professor Siegert and towards the end of this period published the article which led to the adoption of his eponym. In 1933 Hünermann became senior physician and acting director of the Dusseldorf Academic Hospital for Children's and Infectious Diseases, and in 1934 he settled in Koblenz as a consultant paediatrician.

In 1939, at the onset of World War II, Hünermann was conscripted for military service in the German army. He was posted to the eastern front and died at Stalingrad on 29 January 1943.

# Nomenclature

I N 1914, while working in the department of paediatrics, University of Cologne, Conradi[1] published an account of the radiographic and histological manifestations of a condition which he termed "chondrodystrophia fetalis hypoplastica". Thereafter, in 1931, Hünermann introduced the designation "chondrodystrophia calcificans congenita". An early description of the histological features of the condition was attributed to Langhans (1893) by Haynes and Wangner in 1951. In the same year Caffey proposed the name "congenital stippled epiphyses", on a basis of the radiographic appearances. This initiative engendered endless confusion, as stippling of the epiphyses in infancy is a non-specific manifestation of a number of genetic and acquired disorders. The alternative title "chondrodysplasia punctata" gained favour but the recognition of phenotypic and genetic heterogeneity led to further nosological difficulties. The conjoined eponym "Conradi-Hünermann" is now used to define a comparatively mild autosomal dominant form of chondrodysplasia punctata, while additional descriptive or genetic terms are used to denote other types of the disorder.

# References


Caffey J (1951) *Pediatric X-ray diagnosis*. Year Book Publishers, Chicago

Conradi E (1914) Vorzeitiges Auftreten von Knochen und eigenartigen Verkalkungskernen bei Chondrodystrophia fotalis hypoplastica Histologische und Rontgenuntersuchungen. Jahrb Kinderh 80: 86–97

Haynes ER, Wangner WM (1951) Chondroangiopathia calcarea seu punctata: review and case report. Radiology 57: 547–550

Hünermann C (1931) Chondrodystrophia calcificans congenita als abortive Form der Chondrodystrophie. Zschr Kinderh 51: 1–19


[1]Erich Conradi was born on 15 May 1882 in Limbach, near Chemnitz. After qualifying in medicine he worked until 1914 in the Department of Paediatrics, University of Cologne. He then became a paediatrician in Chemnitz for a brief period before commencing military service in World War I. He was severely wounded and died in Bodenweiper on 19 March 1919.

# HUTCHINSON, Sir Jonathan

*(1828–1913)*

*From*: Shelley W.B., Crissey J.T. (1953) Classics in Clinical
Dermatology

*Courtesy*: Charles C. Thomas, Springfield, Illinois
The library of the College of Physicians of Philadelphia

**H**UTCHINSON-GILFORD syndrome, or progeria, is a rare precocious senility disorder in which progressive alopecia, adipose atrophy and deficient growth produce an appearance of old age during childhood. The complications of atherosclerosis lead to premature death, usually before the age of 30 years.

# Biography

**H**UTCHINSON was a Victorian polymath with extensive experience in dermatology, syphilology, ophthalmology, neurology and surgery. He became Professor of Surgery at the London Hospital and he is remembered for his extensive contributions to the medical literature.

Sir Jonathan Hutchinson was born on 23 July 1828 in Selby, Yorkshire, where his father had a prosperous flax business. His family were Quakers and, together with his 11 siblings, he was brought up in a warm, happy, deeply religious home. He shared his parents' religious convictions and initially planned a career as a medical missionary. In 1848 Hutchinson became an apprentice to Caleb Williams, an apothecary and surgeon of York; he completed his medical training and obtained his professional qualification in 1850 at St Bartholomew's Hospital, London. During his student days in London Hutchinson became involved with philanthropic Quaker Missions, with the aim of alleviating misery and uplifting the impoverished.

While still a student, Hutchinson gained the friendship of his mentor, Sir James Paget, whose patronage facilitated his career as a surgeon at the London Hospital and as a member of the staff of Moorfields Eye Hospital and Blackfriars Hospital for Diseases of the Skin. Hutchinson developed a special interest in congenital syphilis, which was common in London in his time, and he was responsible for delineating the natural history of this disorder.

Hutchinson had vast clinical experience and he published his observations in more than 1,200 medical articles. Despite his busy clinical practice he produced the quarterly *Archives of Surgery* from 1890 to 1900; although he was the only contributor, this journal had a wide circulation and was held in high regard. Hutchinson served as president of the British Ophthalmological Society, the Neurological Society and the International Dermatological Congress. He was also elected to the Fellowship of the Royal Society (1882) and was President of the Royal College of Surgeons (1889). Hutchinson received honorary degrees and Fellowships from many universities and societies and in 1908 he was knighted by King Edward VII for his services to medicine.

Although his intellectual attributes were unchallenged, Hutchinson had his critics. Graham Little commented: "He was a tall, stooping, spare figure, wearing an unbeautiful straggling beard at a time when beards were obsolete. He was totally devoid of any sense of humour and like most humourless men, incredibly obstinate in clinging to his opinions long after they had been demonstrated to be untenable".

A more charitable observer, who attended a lecture delivered by Hutchinson, stated: "What we saw that day was a tall man with a great dome of a head, dark eyes looking benevolently through steel-rimmed spectacles, and a white beard which came well down on his chest. He was dressed in a suit of black broad cloth and looked like an absent-minded professor, though there was nothing in the least absent-minded about his delivery. I do not remember what he talked about that day but he held us completely for an hour. He spoke rather slowly and solemnly, and what he said was clear and logical. There was nothing scintillating about it, but you felt he was speaking out of a immense knowledge. Occasionally he illustrated his point by some unexpected simile and there was a distinct north country intonation in his voice which seemed somehow to make what he said more trustworthy!"

In his personal life, Hutchinson had 31 years of happy marriage with his wife, Jane, and their numerous offspring. His family lived in his country home in Haslemere, south of London, which he visited at the weekends and holidays, while he lived in his London house during the working week. He shared the latter establishment with his friends Nettleship, Warren Tay and Hughlings Jackson, all of whom achieved eminence in academic medicine.

Hutchinson held deep moral convictions throughout his life, although he moved away from strict Quaker religious dogma. He died at Hazlemere in 1913, a patriarch aged 85 years, after choosing his own epitaph: "A man of hope and forward-looking mind".

# Nomenclature

**I**N 1866 Hutchinson documented the clinical features of a boy aged 6 years who had congenital absence of the hair and atrophy of the skin. Hastings Gilford, a surgeon at the Reading Dispensary, followed up this patient and another, and in 1904 published a case report terming the condition "progeria" from the Greek word meaning "prematurely old". He provided a set of pictures of the disorder from infancy to the age of 17 years, in which the characteristic alopecia, fat loss and premature aging are clearly depicted.

A possible earlier report in *Phlegontis Tralliani de Mirabilibus* relates to Craterus, who went from infancy to old age and death, within a span of 7 years. Another classical description appeared in the St James Gazette concerning Hopkin Hopkins, a well-known exhibitionist: "March 19th, 1754. Died in Glamorganshire of mere old age and a gradual decay of nature at 17 years and 2 months, Hopkin Hopkins, the little Welshman, lately shown in London. He never weighed more than 17 pounds, but for 3 years past, no more than 12."

The conjoined eponym "Hutchinson-Gilford syndrome" was used by Waldorp and del Castillo (1928) and in 1972 De Busk reviewed 60 cases under this title. Since that time the eponymic form has been supplanted by the descriptive designation "progeria". The genetic basis of progeria is uncertain; although a few instances of affected siblings have been reported, the majority of cases have been sporadic. A paternal age effect has been recognised, and new dominant mutation is a possible mechanism. In addition to progeria, Hutchinson's eponym is applied to several non-genetic disorders which reflect his wide interests: "Hutchinson's incisors", "Hutchinson's pupil", "Hutchinson's potato tumour" and "Hutchinson's freckle".

# References

De Busk FL (1972) The Hutchinson-Gilford progeria syndrome. Report of 4 cases and review of the literature. J Pediatr 80: 697–724

Gilford H (1904) Progeria: a form of senilism. Practitioner 73: 188–217

Hutchinson J (1886) Congenital absence of hair and mammary glands with atrophic condition of the skin and its appendages of a boy whose mother had been almost wholly bald from alopecia areata from the age of six. Trans Med Chir Lon. 69: 473

McKusick VA (1971) Dedication to Jonathan Hutchinson. Part XII Clinical Delineation of Birth Defects, Skin, Hair and Nails. Birth Defects: Original Article Series VII, 7: 1–2

Obituary (1913) Br Med J 1: 1398–1401

Obituary (1913) Br J Dermatol 25: 225–227

Waldorp EP, del Castillo EB (1928) Infantilisme gérodystrophique des auterus: varieté clinique de la "progeria" de Hastings Gilford ou "nanisme sénile" de Variot et Pronneau. Presse Méd 36: 1221–1223

Wales AE (1963) Sir Jonathan Hutchinson. Br J Vener Dis 39: 67–86

# JADASSOHN, Josef

## *(1860–1936)*

J ADASSOHN-LEWANDOWSKI syndrome, or pachyonychia congenita, comprises onychogryposis,
hyperkeratosis of the palms and soles, oral leukokeratosis, hyperhydrosis and premature eruption of the teeth.
The syndrome is an autosomal dominant trait.

# Biography

J ADASSOHN was an outstanding figure amongst
European dermatologists in the early decades of the
twentieth century. During his tenure of professorships in Bern and Breslau he was instrumental in applying
laboratory techniques to the elucidation of skin disorders.

Joseph Jadassohn was born on 10 September 1860 in Liegnitz,
Schlesien and he studied medicine in Göttingen, Heidelberg,
Leipzig and Breslau. He qualified at the University of Breslau
in 1887 and spent the next 5 years training in dermatology as
an assistant to Neisser. In 1896 Jadassohn was called to the
chair of dermatology at Bern and in 1917 he returned to his
alma mater in Breslau, where he was professor of dermatology until his retirement in 1931.

Jadassohn was a fine clinician and a notable teacher. He took
a meticulous scientific approach to his speciality and he was responsible for introducing laboratory methods in the investigation of skin diseases. He was also one of the first to employ
immunological techniques in the study of dermatological disorders and he made special contributions to the understanding
of the immunopathology of tuberculosis and trichophytosis.

Several dermatological disorders were delineated by
Jadassohn and together with his pupils and staff, he published numerous high quality articles. He served as editor
of *Archiv für Dermatologie und Syphilis* and the
*Zentralblatt für Haut- und Geschlechtskrankheiten* and in
1928 he compiled the classical *Handbuch der Haut- und
Geschlechtskrankheiten*.

As a well-known figure in European dermatology
Jadassohn maintained links with colleagues in many coun-
tries. He was a corresponding member of the British
Association of Dermatology and, in the year prior to his
death, he was elected to honorary fellowship of the Royal
Society of Medicine.

Jadassohn was diligent and dedicated, spending long
hours at his work. He was kindly and courteous to his colleagues and staff, whom he inspired with his enthusiasm for
dermatology. He died on 24 March 1936, at the age of 72
years; his son, Dr Werner Jadassohn followed in his foot-
steps in dermatology.

# Nomenclature

J ADASSOHN'S tenure of the chair of dermatology at
the University of Bern was a highly productive period
of his career and in 1906, in collaboration with his
junior colleague, Felix Lewandowski (1879–1921), he documented the disorder which bears their conjoined eponym.
The patient was a girl aged 15 years who was admitted to hospital with fungating tuberculosis of the skin. Abnormalities of
the nails had been present since birth and she also had
unusual keratinisation of the skin and tongue. In their case
description, Jadassohn and Lewandowski stated "The nail
plates of all the fingers and toes are extremely thickened, and
so hard that they cannot be cut with scissors; the father has to
trim them with a hammer and chisel. The fingernails are of
normal length and width, shiny and smooth on the surface,
and show dry whitish streaks here and there at the free edges.
They are translucent on the whole, and only at the very tip of

the fingers are they coloured greyish black. At the distal
parts, especially, they are much more strongly arched trans-
versely than normal; some are also somewhat more curved in
their long axis toward the volar surface. They become thicker
toward the free border, measuring three to five mm. in thick-
ness there. For this reason they project quite markedly on the
dorsal aspect but at the same time they lie on a greatly thickened nail bed; this is, as it were, enveloped by them and com-
pressed from the sides toward the middle so that it stands out
on the free edge as a small addition to the thick, sickle-
shaped plate." Hyperhydrosis of the nose, palms and soles,
scanty papular hyperkeratosis of the knees and elbows and a
white plaque on the tongue were additional features. The
authors mentioned that a younger brother was similarly affected, although the parents and 7 sisters had no skin
problems.

Jadassohn and Lewandowski employed the descriptive title
"pachyonychia congenita" in their article and in the reports
which followed this term was generally used. The aetiology
of the condition remained uncertain but the familial nature
was firmly established in 1921, when Murray documented 7
affected persons in 3 generations of a family. In a similar
report, Kumer and Loos (1935) described 24 cases in a 5-gen-
eration family. The phenotype was expanded and the autosomal dominant mode of inheritance was clinched in 1983 when
Stieglitz and Centerwall published details of a kindred with 17
affected persons in 4 generations. The conjoined eponym fea-
tured in the title of this latter article.

There is scope for semantic confusion as, in addition to
pachyonychia congenita, Jadassohn's name is also attached
to several other conditions, including maculopapular urticaria, naevus sebaceum furunculus atonicus and blue neu-
ronaevus. A further difficulty arises from the fact that familial
pachyonychia congenita may be heterogeneous; this issue
remains unresolved.

# References


Jadassohn J, Lewandowski F (1906) Pachyonychia congenita keratosis dissemi-
nata circumscripta (follicularis). Tylomata. Leukokeratosis linguae. In:
Neisser A, Jacobi E (eds) *Ikonographia Dermatologica*. Urbach &
Schwanzenberg, Berlin, pp. 29–31
Kumer L, Loos HO (1935) Ueber Pachyonychia congenita (Typus Riehl). Wien
Klin Wschr 48: 174–178
Murray FA (1921) Congenital anomalies of the nails. Four cases of hereditary
hypertrophy of the nail bed associated with a history of erupted teeth at
birth. Br J Derm 33: 409–412
Obituary. Br J Derm & Syph 48/6: 323–324
Shelley WB, Grissey JT (1953) Josef Jadassohn. In: Classics in Clinical
Dermatology. Charles C Thomas, Springfield, Illinois
Stieglitz JB, Centerwall WR (1983) Pachyonychia congenita (Jadassohn-
Lewandowsky syndrome): a seventeen-member, four-generation pedigree
with unusual respiratory and dental involvement. Am J Med Genet 14: 21–28

# JERVELL, Anton
## *(1901–1987)*

*Courtesy*: Professor Jak Jervell, Norway

J ERVELL and LANGE-NIELSEN syndrome is an autosomal recessive disorder comprising perceptive deafness and episodes of syncope due to cardiac conduction abnormalities. Electrocardiographs show a characteristic prolongation of the Q-T interval.

# Biography

J ERVELL was a prominent Norwegian physician in the twentieth century. He made important contributions to the development of cardiology in Scandinavia.

Anton Jervell was born in Norway on 14 June 1901 and after obtaining his medical qualification, he specialised in internal medicine. In 1934 Jervell spent a year obtaining postgraduate experience in Paris and he then proceeded to his doctorate with a thesis on the electrocardiographic findings in myocardial infarction. Jervell published extensively on cardiological topics, notably arrhythmias, and he also edited a textbook for nurses. In 1938 he became chief physician and director of the Vestfol County Hospital in Tønsberg, about 70 miles south of Oslo. He held this post until 1956 when he became professor of medicine at Ulleväl Hospital, Oslo, where he remained until his retirement.

In addition to his medical interests, Jervell was active in the Norwegian labour movement and in local politics. During World War II, when Norway was occupied by the Germans, Jervell played a role in the Resistance movement, sheltering refugees in his home and in his hospital beds. In the later stages of his career Jervell was chairman of the Norwegian Medicinal Depot for the distribution of medicines to pharmacies and hospitals.

Jervell was a popular teacher and was held in high esteem by his students and colleagues. He had a gentle bedside manner and was regarded as a role model of a good doctor. He had a happy home life and is fondly remembered by his family and friends for his kindness and hospitality. His son, Jak Jervell, is currently professor of medicine at Oslo University.

# Nomenclature

J ERVELL retained his interest in medicine after his retirement, and in 1983 at the age of 82 years, he addressed a medical society in Oslo on the subject of the condition which bears his name. The content of his presentation was published in 1985, in the last article which he wrote. In this paper he recalled how in 1957, together with his colleague Lange-Nielsen, he investigated a deaf child who had experienced fainting attacks. Electrocardiographic studies revealed a prolonged Q-T interval, which is typical of the disorder. Of the 6 children in the family, 2 other siblings were similarly affected and another had previously died suddenly. Following publication of these observations, Levine and Woodworth of the Peter Bent Brigham Hospital in Boston, reported a boy with similar clinical and electrocardiographic findings. The term "congenital deaf-mutism" was used in the titles of both early papers but in reports which followed this was modified to "cardioauditory" or "surdo-cardiac" syndrome, with the addition of the eponyms of Jervell and Lange-Nielsen.

In 1964 Fraser and colleagues reviewed genetic aspects of the condition and concluded that it was inherited as an autosomal recessive trait. They drew attention to a probable example documented by Meissner in 1856, and provided the following case history: "In Meissner's institution for deaf-mute children a girl had appropriated a trifle from another child and she was called before the director in the presence of the other pupils. As she realized that she was discovered, she was so struck by repentance and grief, that in an instant she sank dead to the ground. The tragic death appeared to the gathered pupils as an act of God as a punishment for her transgression, reminding them never to stray from the path of virtue. When the parents were informed of her unhappy end, they were not at all surprised. On the contrary, they replied that the news had not come as a shock since similar events had previously occurred in their family. It transpired that another deaf-mute child had died suddenly after a terrible fright, and a third after a violent fit of rage".

Romano and colleagues (1963) documented an Italian family with prolongation of the Q-T interval and a propensity to syncope, and in the following year Ward reported an Irish family with the same disorder. This condition, which is now termed the "Romano-Ward syndrome" does not have deafness as a syndromic component, and the autonomous syndromic identity of the Jervell and Lange-Nielsen syndrome remains intact.

# References


Fraser GR, Froggat F, Murphy T (1964) Genetical aspects of the cardio-auditory syndrome of Jervell and Lange-Nielsen (congenital deafness and electrocardiographic abnormalities). Ann Hum Genet 28: 133–137

Jervell A (1985) The surdo-cardiac syndrome. European Heart J(Suppl D) 6: 97–102

Jervell A, Lange-Nielsen F (1957) Congenital deaf-mutism, functional heart disease with prolongation of the Q-T interval, and sudden death. Am Heart J 54: 59–68

Levine SA, Woodworth CR (1958) Congenital deaf-mutism, prolonged Q-T interval, syncopal attacks and sudden death. N Engl J Med 259: 412–417

Meissner FL (1856) Taubstummheit und Taubstrummenbildung. Leipzig u. Heidelberg p. 119

Romano C, Gemme G, Pongiglione R (1963) Aritmie cardiache rare dell'età pediatrica. Clin Pediat Bologna 45: 680–683

Ward OC (1964) A new familial cardiac syndrome in children. J Irish Med Ass 54: 103–106

# JEUNE, Mathis

## *(1910–1983)*

*Courtesy*: Professor Henri Plauchu, Lyon

J EUNE syndrome or asphyxiating thoracic dystrophy presents in the newborn with marked constriction of the rib cage and specific radiological changes. Polydactyly is a feature in some affected infants and chronic nephritis is a late complication in survivors. Inheritance is autosomal recessive.

# Biography

J EUNE was a prominent academic paediatrician who enjoyed a successful career in Lyon, France, during the twentieth century.

Mathis Jeune was born in France in 1910. He was a brilliant student, qualifying in medicine at the University of Lyon and becoming "interne lauréat" at the age of 23 years. He initially trained in adult internal medicine but eventually became interested in paediatrics and by 1947 he had obtained a senior position in this speciality at the Hôpital St Eugénie, Lyon. He moved to the Hôpital Debrousse in 1951 and in 1964 he achieved the status of Professor of Clinical Paediatrics.

Jeune made many academic contributions over a 30-year period; he was held in high regard by his students, and was admired for his moral authority and his intellectual qualities. His early research was concerned with lung disease and the pulmonary physiology of infancy and he made significant observations on the bronchial complications of primary tuberculous infection. He subsequently worked in the field of metabolic disorders and he developed a special interest in phenylketonuria. In the last 15 years of his hospital life, Jeune was oriented towards the endocrinology of infants and he made notable contributions to the understanding of sexual ambiguity and thyroid dysfunction.

In addition to his academic accomplishments, Jeune was deeply involved in the social aspects of paediatrics and he held office in several organisations which were dedicated to this field. Jeune also served as President of a number of local medical associations and as editor of the journal *Pédiatrie*.

After Jeune's death in Lyon in 1983, at the age of 73 years, his colleague Professor Larbre gave a eulogy in his honour at a meeting of the Paediatric Society of the Rhone Alpes region.

# Nomenclature

I N 1955, in collaboration with his colleagues Beraud and Carron, Jeune documented 2 infants with long narrow thoraces and short limbs who died from respiratory distress. The affected neonates were related, and in the title of their article "asphyxiating thoracic dystrophy" Jeune and his colleagues drew attention to the familial nature of the disorder. Prior to this report, Jeune had published his observations in 1954 but the second publication is usually quoted in the literature.

The familial nature of the condition was emphasised in 1966 when Pirnar and Neuhauser reported 3 affected brothers, and 2 further sets of siblings with the disorder were documented in the following year by Hanissian et al. (1967). These latter authors drew attention to the similarity of the radiographic changes to those of the Ellis-van Creveld syndrome. Autosomal recessive inheritance was established by Shokeir et al. (1971) in an account of an affected Norwegian kindred in which minor thoracic abnormalities were a possible indicator of heterozygosity.

By 1995, more than 100 affected infants had been reported and the range of clinical manifestations had been widened to include late onset glomerulonephritis and hepatic dysfunction with cirrhosis. The alternative anatomical designation "thoracic-pelvic-phalangeal dystrophy" has been introduced, but the original style "asphyxiating thoracic dystrophy" or the eponymous form "Jeune's syndrome" is generally preferred.

In addition to asphyxiating thoracic dystrophy, Jeune's name is associated with that of Tommasi in the title of an autosomal recessive syndrome which comprises ataxia, deafness and cardiomyopathy. This disorder, which Jeune and his colleague delineated in 1963, is extremely rare.

# References

Hanissian AS, Riggs WW Jr, Thomas DA (1967) Infantile thoracic dystrophy – a variant of Ellis-van Creveld syndrome. J Pediat 71: 855–864

Jeune M, Carron R, Beraud C, Loaec Y (1954) Polychondrodystrophie avec blocage thoracique d'évolution fatale. Pediatrie, Lyon 9(4): 390–392

Jeune M, Beraud C, Carron R (1955) Dystrophie thoracique asphyxiante de caractère familial. Arch Fr Pediatrie 12: 886–891

Jeune M, Tommasi M, Freycon F (1963) Syndrome familial associant ataxie, surdité et oligophrénie. Sclérose myocardique d'évolution fatale chez l'un des enfants. Pédiatrie, 18: 984–987

Larbre (1983) Allocution prononcée par le Professor Larbre à la Réunion Pédiatrique de la Région Rhône-Alpes, lors de sa séance du 27 septembre 1983

Pirnar T, Neuhauser EBD (1966) Asphyxiating thoracic dystrophy of the newborn. Am J Roentgen 98: 358–364

Shokeir MHK (1970) Asphyxiating thoracic chondrodystrophy: association with urinary malformations and evidence for heterozygous expression. (Abstract). Am J Hum Genet 22: 18A–19A

Shokeir MHK, Houston CS, Awen CF (1971) Asphyxiating thoracic chondrodystrophy: association with renal disease and evidence for possible heterozygous expression. J Med Genet 8: 107–112

# KALLMANN, Franz Josef
*(1897–1965)*

K ALLMANN syndrome is a genetically heterogeneous disorder in which hypogonadism and anosmia are the major manifestations.

# Biography

K ALLMANN was one of the first people to study the genetic basis of psychiatric disorders. He spent most of his career in New York, where he pioneered the use of twin studies in the assessment of the relative roles of heredity and the environment in the pathogenesis of psychiatric disease.

Franz J. Kallmann was born in 1897 in Silesia, formerly in Germany, now Poland, where his father was a physician and surgeon. He qualified in medicine at the University of Breslau in 1919, being taught for a time by Alzheimer, and undertook internship at the Allerheiligen Hospital in that city. After a few years in private practice, he trained in psychiatry in Berlin and Munich. It was at this stage of his career that he developed his lifelong interest in the genetic basis of schizophrenia. In 1929 Kallmann became director of two state hospitals in Berlin. With the advent of the Nazi era, compulsory sterilization of psychotic patients was introduced. Kallmann's opposition to eugenic measures such as this led to him being banned from publishing or speaking at medical meetings. His problems were compounded by the fact that although his father had converted from Judaism to Christianity, the Nazis regarded him as being Jewish. Pressures increased and in 1936 Kallmann moved to New York.

His early years in North America were difficult, partly because of language complications, but Kallmann eventually became chief of psychiatric research in medical genetics at the New York State Psychiatric Institute and head of medical genetics at the Columbia-Presbyterian medical centre. At the time, Kallmann's department was unique by virtue of its focus on the genetics of psychiatric disorders, and his group was pre-eminent in this field. Sets of twins formed the basis of his investigations and he used this approach for studies of psychoses, mental deficiency, aging and longevity.

While in Germany Kallmann had written a treatise entitled *The Genetics of Schizophrenia*. He brought the manuscript with him to New York, translated it into English and published it in 1938. His academic output continued and in 1953 he was the author of *Heredity in Health and Mental Disorder*. Kallmann subsequently edited *Expanding Goals in Psychiatry* (1962) and *Family and Mental Health Problems in a Deaf Population* (1963). In addition to these monographs, Kallmann was the author of more than 150 medical articles. His work made an impact amongst medical geneticists but many of his psychiatric colleagues rejected the notion that differences in human behaviour could have a genetic basis. These difficulties only ameliorated towards the end of his career and, in some instances, they still remain unresolved.

Kallmann's achievements in both psychiatry and genetics were recognised by his election as president of the American Psychopathological Association, chairman of the Permanent Committee for International Congresses of Human Genetics and president of the American Society of Human Genetics. In the latter role, in 1952, he made strenuous and successful efforts to ensure a viable membership. He also made a major contribution to the establishment of the Society's journal *American Journal of Human Genetics*.

Kallmann developed a chronic, relapsing illness and he died in 1965 at the age of 67 years in the Harkness Pavilion of the Colombia-Presbyterian medical centre. In 1966 his widow, Helen (Helly), received the posthumous Stanley R. Dean award of the University of Michigan for his contributions to the understanding of schizophrenia. In a eulogy, his colleague John D. Rainer stated "Franz Kallmann was a many-sided person. He was a scientist in the broadest sense with a fertile imagination, a thorough knowledge of subject matter and method, a scanning interest in all human activity and the constant ability to frame richly suggestive hypotheses and to formulate careful research plans for their investigation. At the same time, he was always a good physician, a knower of men and a student of human fortitude and weakness, a family counsellor and a clinical psychiatrist in the noblest tradition."

# Nomenclature

I N 1856, Maestre de San Juan, a Spanish pathologist, described the autopsy finding in a man with absent olfactory lobes and poorly developed genitalia. Thereafter Weidenreich (1914) undertook post mortem studies in 10 persons with anosmia, 3 of whom had hypogonadism, and suggested that these problems might be syndromically associated. Thirty years later, in 1944 Kallmann and his colleagues reported the combination of hypogonadism and anosmia in members of 3 separate families and drew attention to the genetic aetiology of the disorder.

In 1954 De Morsier documented 14 affected persons and confirmed that the anosmia was the consequence of agenesis of the olfactory lobes; for a time the title "dysplasia olfactogenitalis of De Morsier" enjoyed some favour but the eponym "Kallmann" is now well established. The alternative descriptive term "olfactogenital dysplasia" is sometimes employed.

The phenotype has now been widened to include facial clefting and other abnormalities and genetic heterogeneity has been recognised. Dominant, recessive and X-linked forms have been defined and progress has been made towards the identification of the chromosomal loci of the underlying gene defects.

# References

De Morsier G (1954) Études sur les dysraphies cranio-encéphaliques. 1. Agénésie des lobes olfactifs (télencéphaloschizis latéral) et des commissures calleuse et antérieure (télencéphaloschizis médian): la dysplasie olfacto-génitale. Schweiz Arch Neurol Neurochir Psychiatr 74: 309–361

Kallmann FJ, Schoenfeld WA, Barrera SE (1944) The genetic aspects of primary eunuchoidism. Am J Ment Defic 48: 203–236

Maestre de San Juan A (1856) Teratologia: falta total de los nervios olfactorios con anosmia en un individuo en quien existia un atrofia congenita de los testiculos y miembro viril. El Siglo Medico 2: 211

Obituary (1965) Br Med J 1: 1440

Weidenreich F (1914) Über partiellen Riechlappendefect und Eunuchoidismus beim Menschen. Z Morphol Anthropol 18: 157

White BJ, Rogol AD, Brown KS, Lieblich JM, Rosen SW (1983) The syndrome of anosmia with hypogonadotropic hypogonadism: a genetic study of 18 new families and a review. Am J Med Genet 15: 417–435

# KIRNER, Joseph
## *(1888–1964)*

*Courtesy*: Professor H.-R. Wiedemann, Kiel

**K**IRNER anomaly, or dystelephalangy, comprises flexion and palmar deviation of the fifth fingers. The condition is an autosomal dominant trait.

# Biography

**K**IRNER was chief physician in a provincial hospital in Germany during the first half of the twentieth century.

Joseph Kirner was born on 13 November 1888 in Augsburg. He qualified in medicine in 1913 and served as a medical officer in the German Army during World War I, receiving a decoration for gallantry. After the war, Kirner joined the staff of the Rheydt Hospital and in 1926 he became senior physician at the Waldshut Hospital, Baden, Southern Germany, where he spent the rest of his career.

Kirner was a kindly, diligent and dedicated physician with medical experience and skills which embraced surgery, gynaecology and radiology. His opportunities for academic pursuits were limited and he published only a few medical articles. Kirner's main contributions and achievements centred around the management and development of the Waldshut Hospital, which had a history dating back to 1411.

Kirner had exceptional personal qualities which were outlined in his obituary: "He was a warm person who emitted an air of trust and respect. He was always working extremely hard, and offered his services selflessly to his hospital and his patients. He was always available for everyone, rich or poor. He practised medicine because he considered it his calling, and because of concern for others. He created a wonderful atmosphere at his hospital in Waldshut. People followed him and it was not necessary for him to issue commands."

Kirner retired in 1959 at the age of 70 years and died suddenly in 1964. He is commemorated by a brass mural relief which has been placed at the entrance to the Waldshut Hospital.

# Nomenclature

**I**N 1927, shortly after he had joined the staff of the Waldshut Hospital, Kirner published an account of the abnormality which bears his name. Wilson published further examples in 1952 and documented generation to generation transmission. In an article from Japan, Sugiura and colleagues (1961) proposed the descriptive designation "dystelephalangy" and the eponymous format "Kirner's deformity" was used by Blank and Girdany in 1965, in an account of an affected family. In 1972 David and Burwood undertook a population survey and identified 18 affected persons in 9 families. Pedigree analysis was suggestive of autosomal dominant inheritance with incomplete penetrance. No systemic ramifications or additional abnormalities were identified in the affected persons.

The terms "camptodactyly" and "clinodactyly" are often used loosely to embrace the Kirner anomaly. In the specific sense "camptodactyly" denotes flexion contractures at the proximal interphalangeal joints, while "clinodactyly" implies radial curvature, usually of the fifth finger.

# References

Blank E, Girdany BR (1965) Symmetric bowing of the terminal phalanges of the fifth fingers in a family (Kirner's deformity). Am J Roentgen 93: 367–373

David TJ, Burwood RL (1972) The nature and inheritance of Kirner's deformity. J Med Genet 9: 430–433

Kirner J (1927) Doppelseitige Verkrümmungen des Kleinfingerendgliedes als selbständiges Krankheitsbild. Fortschr Geb Röntgen 36: 804–806

Sugiura Y, Ueda T, Umezawa K, Tajima Y, Sugiura I (1961) Dystelephalangy of the fifth finger: dystrophy of the fifth finger. J Jpn Orthop Assoc 34: 29–35

Wilson JN (1952) Dystrophy of fifth finger: report of four cases. J Bone Joint Surg 34B: 236–239

# KORSAKOV, Sergei Sergeivich

*(1854–1900)*

*From*: Katzenelbogen S. (1953) In: Webb Haymaker (ed) The Founders of Neurology

*Courtesy*: Charles C. Thomas, Springfield, Illinois
Dr Maurice Genty, Académie Nationale de Médecine, Paris

K ORSAKOV psychosis comprises amnesia and confabulation, frequently in association with polyneuritis. The condition, which occurs principally in alcoholics, is the consequence of an autosomal recessive defect of thiamine metabolism, which becomes apparent when the dietary intake is deficient.

# Biography

K ORSAKOV was the founder of modern Russian psychiatry. In addition to his academic contributions, he is remembered for his humanitarianism.

Sergei Sergeivich Korsakov was born on 22 June 1854 on the Gus Estate, Vladimir Province, Russia, where his father was the manager of a glass-making factory. He qualified in medicine at the University of Moscow in 1875 and became a physician at the Preobrazhenskii mental hospital. Korsakov gained postgraduate experience in the clinic for nervous diseases, under Professor Kozhevnikov and obtained a doctorate in 1887 on the subject of "alcoholic paralysis". In the years that followed Korsakov succeeded to the Chair of Psychiatry and became superintendent of the new Moscow University Psychiatric Clinic.

Korsakov had ample opportunity to study the effects of alcoholism and early in his career he received wide recognition for his descriptions of the disorder which bears his name. He was the first to give a clear account of paranoia and he made many other contributions, including a classification of mental disorders, and an analysis of the role of social factors in psychiatric disease. He published a textbook of psychiatry in the Russian language and he was the author of many articles. Korsakov was a humanitarian and he was ahead of his time in implementing a policy of freeing his psychotic patients from physical restraints and for attempting rehabilitation within a family framework. He also concerned himself with the welfare of his students, attempting to provide financial support for those in need.

Korsakov founded the Moscow Society of Neuropathologists and Psychiatrists in 1890 and in 1897 he organised the 12th International Medical Congress, which was held in Moscow. He then worked out a constitution for a proposed Russian Association of Psychiatrists and Neurologists, which was successfully established after his death.

Korsakov died on 1 May 1900 at the age of 46 years. A phrase, written when he was a boy aged 11 years, sums up his philosophy – "Help others. When the occasion presents itself to do good deeds, do them. Withdraw from evil."

# Nomenclature

A N early mention of the condition which bears Korsakov's name was provided in 1822 by Dr James Jackson of the Massachusetts General Hospital, Boston, in a review of the peripheral neuritis of alcoholism. Subsequently, in 1868, Sir Samuel Wilks, physician at Guy's Hospital, London, gave an account of the characteristic mental symptoms in an article on alcoholic paraplegia.

Korsakov (sometimes spelt "Korsakoff") gave details of the disorder in his doctoral thesis of 1887, using the title "cerebropathia psychica toxaemica". In the same year he wrote an account in the Russian journal *Vestnik Psychiatrie* and this was followed by further articles in the German literature. Korsakov's publications were translated and reviewed by Victor and Yakovlev in 1955. Korsakov emphasised the association of alcoholic polyneuropathy with a specific pattern of mental disturbance as follows "This mental disorder appears at times in the form of sharply delineated irritable weakness of the mental sphere, at times in the form of confusion with characteristic mistakes in orientation for place, time and situation, and at times as an almost pure form of acute amnesia, where the recent memory is most severely involved, while the remote memory is well preserved .... Some have suffered so widespread memory loss that they literally forget everything immediately." Korsakov subsequently pointed out that the mental abnormalities could occur in the absence of polyneuropathy and that this type of amnesia could result from conditions other than alcoholism.

Following publication of Korsakov's papers the eponym was soon established. The term "Korsakov's psychosis" was used for the combination of alcoholic polyneuropathy and characteristic mental disturbance, while "Korsakov syndrome" was reserved for the non-alcoholic forms, with or without polyneuritis, which may result from acquired conditions, such as head injury, brain tumours and encephalitis.

The natural history of the disorder in 245 affected persons, in 82 of whom autopsy was undertaken, was documented by Victor et al. (1971). A genetically determined disturbance of transketolase, a thiamine pyrophosphate binding factor was identified in fibroblasts from affected persons, by Blass and Gibson (1977). It is now assumed that persons who are homozygous for this defect are at increased risk of developing thiamine deficiency in circumstances of dietary inadequacy, notably alcoholism. In this way Korsakov psychosis can be regarded as an autosomal recessive disorder which manifests in genetically predisposed persons in whom alcoholism leads to thiamine deficiency. Wernicke encephalopathy, an alcoholism-related confusional state associated with oculomotor paralysis, nystagmus and ataxia, is now known to result from a similar genetic predisposition. In strict terminology, Korsakoff psychosis is a specific pattern of amnesia and confabulation, while Wernicke encephalopathy is characterised by confusion and the above-mentioned neurological disturbance. If both are present in the same person, the conjoined eponym "Korsakov-Wernicke" is employed (*see* Wernicke, p. 191).

# References


Blass JP, Gibson GE (1977) Abnormality of a thiamine-requiring enzyme in patients with Wernicke-Korsakoff syndrome New Engl J Med 297: 1367–1370

Korsakoff SS (1887) Disturbance of psychic function in alcoholic paralysis and its relation to the disturbance of the psychic sphere in multiple neuritis of nonalcoholic origin. Vestnik Psichiatrii Vol IV/2

Korsakoff SS (1889) Psychic disorder in conjunction with multiple neuritis (Psychosis Polyneuritica S Cerebropathia Psychica Toxaemica). Medizinskoje Obozrenije 31, No. 13

Korsakoff SS (1890) Über eine besondere Form psychischer Störung kombiniert mit multipler Neuritis. Arch Psychiat Nervenkr 21

Korsakoff SS (1891) Über Erinnerungsstörungen (Pseudoreminiszenzen) bei polyneuritischen Psychosen. Allg Z Psychiat 47

Victor M, Adams RD, Collins GH (1971) The Wernicke-Korsakoff syndrome: a clinical and pathological study of 245 patients, 82 with post-mortem examinations. FA Davis, Philadelphia

Victor M, Yakovlev PI (1955) SS Korsakoff's psychic disorder in conjunction with peripheral neuritis. A translation of Korsakoff's original article with brief comments on the author and his contribution to clinical medicine. Neurology (Minneap) 5: 394–406

Wilks Sir S (1868) Alcoholic Paraplegia. Medical Times and Gazette 2: 467–472

# LITTLE, William John

## *(1810–1894)*

L ITTLE disease, or cerebral palsy, is a common form of neurological disturbance which usually results from brain damage sustained in the perinatal period. There is considerable aetiological heterogeneity and genetic factors are occasionally implicated.

# Biography

L ITTLE was a London physician during the nineteenth century. He is best remembered for his description of the pathogenetic factors involved in the condition which bears his name; it is of historical interest that he suffered from a left talipes equinovarus deformity.

William John Little was born in 1810 at the Red Lion Inn, Aldgate, in the East End of London, where his father was the landlord. He was handicapped by a deformed left foot which might have been the consequence of poliomyelitis contracted during early childhood. He went to school in Dover, Kent, until the age of 13 years and thereafter at the famous Jesuit College of St Omer, France. Little had considerable linguistic ability and gained a prize for French composition prior to his return to England in 1826. He then spent 2 years apprenticed to an apothecary before commencing medical studies at the London Hospital in 1828. After qualification in 1831 Little gained further experience in anatomy, pathology and physiology at Guy's Hospital and University College, before being appointed as lecturer in these subjects at the London Hospital Medical School in 1836. An unsuccessful application for a post as surgeon at the London Hospital prompted Little to change the direction of his career and he moved to Berlin in order to acquire the necessary postgraduate experience for the licence of the Royal College of Physicians.

Little's choice of Berlin was doubtless influenced by the work of a German surgeon, Stromeyer, of Hanover, who had pioneered tenotomy in the operative management of clubfoot. After much heart searching Little submitted himself to this operation and, to his great joy, his disability was much improved. He became an enthusiastic proponent of the tenotomy operation and treated more than 30 patients in Berlin. On his return to London in 1837 he used this experience, together with the results of his histopathological investigations in Berlin, for his doctoral dissertation on the pathogenesis and management of clubfoot. Little settled into practice in Finsbury Square, where he published a treatise on *Clubfoot and Analogous Distortions*. This work attracted local and international attention, with the result that Little was able to initiate the foundation of the Orthopaedic Institution, London. This became the Royal Orthopaedic Hospital and later evolved into the present day Royal National Orthopaedic Hospital.

He re-established his links with the London Hospital and in 1845 was appointed to the staff as a physician. Little occupied this post until 1863 when he resigned in order to concentrate upon his private practice. The hospital governors commented on "his great professional ability, his unvarying kindness towards his patients and his courteous behaviour on all occasions throughout a lengthened and highly valued connection of more than 23 years with this important charity". In addition to his practice, Little held honorary appointments as visiting physician to the National Orthopaedic Hospital, the Asylum for Idiots at Reigate, the Infant Orphan Asylum, and the Royal Hospital for Incurables.

Little made a number of important contributions to the orthopaedic literature. In 1843 he published a treatise on *Ankylosis or Stiff Joints* and in 1868 another on *Spinal Weakness and Spinal Curvatures*. Little also wrote *Treatment of Deformities in the Human Frame* published in 1853, which contained descriptions of 2 boys with "pseudohypertrophic paralysis"; several years later Duchenne documented further examples in Paris and thereafter his own name was attached to this form of muscular dystrophy. In 1882, in collaboration with one of his two medically qualified sons, Little wrote a monograph on the management of knock-knees.

By 1884 Little was becoming deaf and he retired to a country home in Walthamstow. He subsequently moved to West Malling, Kent, where he died in 1894 at the age of 84 years after a short illness. In 1981 a plaque was erected on the Old Red Lion Inn, commemorating Little's birth, unveiled by his grandson, Admiral Sir Charles Little.

# Nomenclature

L ITTLE'S academic interests were influenced to a significant extent by his own foot deformity. His private practice and his various hospital appointments brought him into contact with large numbers of children with various neuromuscular and articular abnormalities, including many with the sequelae of cerebral palsy. Little realised that a variety of complications during parturition, some of which were potentially preventable, could lead to brain damage and spasticity. He eventually crystalised his experience of more than 20 years in a paper which he read to the Obstetrical Society in 1861 and which was published in the following year. In this presentation Little mentioned that he had examined more than 200 patients with "spastic rigidity excluding those in idiot and other asylums, where such cases abound". He went on to mention the clear clinicopathological correlation between brain abnormalities detected at autopsy and spastic paralysis, and he listed the causative factors, emphasising that asphyxia neonatorum was by far the most common. He then gave a detailed review of the clinical manifestations and natural history of the condition, emphasising the presence of spastic rigidity.

Little's paper was the first real account of the pathogenesis of spasticity in the newborn. It was received with enthusiasm and his name soon became identified with cerebral palsy, this format receiving especial acceptance in the USA. Other descriptive synonyms which have been used include cerebral diplegia, cerebral spastic paralysis, congenital spastic diplegia and infantile spastic diplegia. Although Little emphasized causal heterogeneity, the concept of a possible genetic influence was only propagated long after his death. It has now become abundantly evident that amongst the large numbers of patients with the wide non-specific diagnosis of "cerebral palsy" or Little disease, there are a few in whom the brain damage is the consequence of a wide variety of Mendelian disorders.

It is noteworthy that Little's original observations on cerebral palsy are largely in keeping with modern concepts and he richly deserves eponymous immortality for his description of the manifestations and natural history of this common disorder.

# References

Before Our Time (1961) William John Little 1810–1894. Lancet 2: 596

Little WJ (1862) On the influence of abnormal parturition, difficult labours, premature birth and asphyxia neonatorum, on the mental and physical conditions of the child, especially in relation to deformities. Tr Obst Soc London 3: 293–344

Obituary (1894) Br Med J, 14 July: 106

Obituary (1894) Lancet 21 July: 168–169

# LOBSTEIN, Jean Fréderic

*(1777–1835)*

*Courtesy*: Mr D. Stewart, Royal Society of Medicine, London

LOBSTEIN disease, now known as osteogenesis imperfecta tarda, or OI type I, is an autosomal dominant disorder characterised by blue sclerae and frequent fractures due to skeletal fragility (*see* van der Hoeve, p. 173, Vrolik, p. 183).

# Biography

LOBSTEIN occupied the first professorial chair of pathology ever to be established and he ranks amongst the founders of this speciality.

Jean Fréderic Lobstein was born in 1777 in Giessen (Hesse), where his father was a university teacher and a minister of the Protestant church. The family returned to their home town of Strasbourg, Alsace in 1790 at the beginning of the French Revolution. This proved to be a tragic move as Lobstein senior was subsequently imprisoned for his sympathy with the Royalist cause and died 9 months later. After the death of his father, Jean Fréderic Lobstein, at the age of 17 years, was left with the obligation of maintaining his widowed mother and his 4 younger brothers. At this time he had commenced medical studies and he joined the Army of the Rhine with the rank of surgeon, third class. After 9 years of active service Lobstein left the army and returned to Strasbourg, where he graduated as doctor of medicine with a thesis entitled "On the nutrition of the fetus". He then received an appointment as prosector in the Faculty of Anatomy and later became head of obstetrics at the Civil Hospital.

Lobstein firmly believed in the value of anatomico-pathological specimens in the teaching of medicine and in 1813 he founded a pathological museum. This initiative was well received and attracted numerous academic visitors. It also brought credit to Lobstein, and counted towards his appointment in 1819 to a newly established Chair of Pathological Anatomy at the University of Strasbourg. This was the first independent chair of pathology ever to be created, and its inception represented a landmark in the development of that speciality. He introduced chemical pathology into his new department and promoted the use of autopsy as a teaching medium. He also devoted himself to the development of his museum, which eventually contained more than 3,000 specimens. In 1824 Lobstein expanded his activities, accepting the chair of clinical medicine and continuing in both fields for the remainder of his career.

In addition to many publications, Lobstein's contribution was a 4-volume work on pathological anatomy, based upon his vast personal experience. In this treatise he introduced the term "arteriosclerosis", which is still widely used. He also alluded to the adult form of the brittle bone disorder which is now known as osteogenesis imperfecta tarda (vide infra).

Outside his medical career Lobstein was active in the study of history and archaeology. He had a keen interest in the excavation of artifacts, which he preserved in his personal museum. He was also a noted numismatist and amassed an extensive collection of coins and medallions. Lobstein was held in high regard by his colleagues and his death from an urinary infection in 1835 was marked by numerous eulogies.

Lobstein's biography, upon which this account is based, was written by Alexander Brunschwig of Chicago, following his own fellowship in medicine at Strasbourg in the early 1930s. Brunschwig pointed out that Strasbourg had been annexed by the German Empire in 1870, as part of Alsace-Lorrain and that the Strasbourg medical faculty had then become part of the Kaiser Wilhelm Hospital. A new institute of pathology was established, with Von Recklinghausen as director, and Lobstein's museum specimens were displaced or dispersed. Strasbourg was ceded back to France after World War I and the name of the University Hospital was changed to that of "Louis Pasteur". A tablet inscribed in Lobstein's memory remains, and his marble bust, created in 1855 by Philip Gross (1801–1876), retains pride of place in an exhibit in the council room of the Faculty of Medicine.

# Nomenclature

THE early history of osteogenesis imperfecta (OI) has been documented in successive editions of McKusick's classical monograph *Heritable Disorders of Connective Tissue*. In his review McKusick mentioned that the earliest example of the condition is the skeleton of an Egyptian mummy dating from about 1,000 BC, which is preserved in the British Museum. An abnormal femur from an Anglo-Saxon burial ground of the seventh century might also represent the disorder. There is also evidence to suggest that Ivor the Boneless, who led the Viking invasion of England in the ninth century, could have been affected. Two centuries later, Ivor's skeleton was exhumed and burnt by William the Conqueror, so objective diagnostic criteria no longer exist.

The first clear description of OI was provided by Ekman in 1788, in a doctoral thesis for the University of Uppsala, in which he described a family in which persons in 3 generations had a condition which he termed "osteomalacia congenita". In 1831 Edmund Axmann of Westheim, Germany, documented the condition in himself and 2 brothers, mentioning that they had blue sclerae and a tendency to dislocations and fractures. In his textbook of pathological anatomy, published in 1835, Lobstein mentioned the adult form of OI, terming the condition "osteopsathyrosis idiopathica". His successors at Strasbourg maintained interest in the disorder and in 1889 Stilling published an account of the histopathological features, while in 1897 Schmidt suggested that the newborn and adult forms represented the same entity.

The eponym "Lobstein" was often employed in reports of the disorder which followed the publication of his treatise but by the beginning of the present century the descriptive term "osteogenesis imperfecta" was in general use. With the subdivision into the "tarda" and "congenita" forms of OI, there was a tendency to use Lobstein's name for the former and that of Vrolik for the latter. The eponym of van der Hoeve has also been attached to OI tarda. With the recognition of further heterogeneity these eponyms have become redundant although the eponymic form "maladie de Lobstein" has continued to appear in the French medical literature, it is now falling into disuse.

(For further details of the evolution of the nomenclature, *see* Vrolik p. 183 and van der Hoeve, p. 173.)

# References


Axmann E (1831) Merkwürdige Fragilität der Knochen ohne dyskrasche Ursache als krankhafte Eigenthümlichkeit drier Geschwister. Ann Ges Heilk (Karlsruhe) 4: 58

Brunschwig A (1933) Jean Fréderic Lobstein, the first professor of pathology. Ann Med Hist 5: 82–84

Ekman OJ (1788) Dissertatio medica descriptionem et casus aliqot osteomalacia sistens. Uppsala, Sweden

Lobstein JG (1835) *Lehrbuch der pathologischen Anatomie*. Stuttgart 179 pp

Obituary (1835) Caillot. Mort de Lobstein. Arch Med de Strasbourg 1: 152

Schmidt MB (1897) Die allgemeinen Entwicklungshemmungen der Knochen. Ergebn Allg Path 4: 612

Stilling H (1889) Osteogenesis imperfecta. Virchows Arch 115: 357

# LOUIS-BAR, Denise
## *(b. 1914)*

*From*: Evrard P., Beya A., Provis M. (1990) In: Ashwal S.
(ed) The Founders of Child Neurology

*Courtesy*: Dr Philippe Evrard, Boston

L OUIS-BAR syndrome, or ataxia-telangiectasia, comprises progressive neurological degeneration, telangiectasia of the skin and mucous membranes, and a propensity to infection due to immunological deficiency. Inheritance is autosomal recessive.

# Biography

L OUIS-BAR practised in Belgium as a psychiatrist, with a special interest in the care of the mentally handicapped.

Denise Bar was born in 1914 in Liège, Belgium but spent the first decade of her life with her parents in Spain. The family returned to Belgium when she was 10 years of age and, after completing her schooling, she entered the Free University of Brussels, where she received her medical qualification in 1939. Concurrently, she obtained a degree in the contemporary equivalent of sports science.

Shortly after qualification Bar married F. Louis, a civil engineer, and thereafter used the hyphenated surname Louis-Bar. Her husband joined the Belgian army in the Ardennes at the onset of World War II and Louis-Bar became a resident at the Bunge Institute of Neurology, Antwerp, where she gained experience in this speciality under Ludo van Bogaert (*see* p. 171). In 1943 she was appointed as instructor in pharmacology at the University of Liège, and in 1945 she became a neuropsychiatrist in the department of internal medicine. Louis-Bar remained in this post until 1957, when she moved with her husband to Brussels on his elevation to headship of the Belgian national office for nuclear energy.

Louis-Bar spent the remainder of her career in Brussels, where she devoted herself to persons with mental handicap. In particular, she made major contributions to the establishment of special facilities for their care and to the development of techniques for their medical and social management.

# Nomenclature

I N 1940 while undertaking an outpatient clinic at the Bunge Institute, Antwerp, Louis-Bar examined a girl aged 9 years in whom progressive ataxia, mental retardation and telangiectasia had developed in early childhood. Her mentor, van Bogaert, suggested that she should document this patient and her findings were published in the following year. In her article, which was written in the French language, Louis-Bar concluded that the condition was an undelineated phakomatosis.

Louis-Bar's paper appeared during World War II and the condition did not receive wide attention at that time. In 1958 Boder and Sedgwick reported the familial nature of the disorder, mentioned Louis-Bar's original paper and acknowledged a personal communication from van Bogaert. Interest was now aroused and in 1960 Pelc and Vis proposed that the condition should be termed the "Louis-Bar syndrome"; many reports followed, and by 1963 Boder and Sedgwick were able to review 101 cases. The involvement of IgA deficiency and a propensity to malignancy were recognised, and the condition was established as an autosomal recessive disorder. At the present time the descriptive title "ataxia telangiectasia" is generally preferred to the eponymous format "Louis-Bar syndrome".

With hindsight, it has become apparent that the condition which carries Louis-Bar's name had been previously reported in 1926 in the French literature by two Czech colleagues, Syllada and Henner. In their article they documented the presence of athetosis and conjunctival telangiectasia and

recognised that the disorder was familial. This early report did not come to medical notice until several decades later, and it seems highly probable that Louis-Bar and van Bogaert were unaware of this Czech priority.

There are interesting historical parallels in the documentation of the Louis-Bar and the Hurler syndromes. The original descriptions of both these disorders were written by young female doctors in the early stages of their careers, their articles appeared during World War II and World War I respectively, and both were published in the Continental literature (French and German).

# References


Boder E, Sedgwick RP (1958) Ataxia-telangiectasia: a familial syndrome of progressive cerebellar ataxia, oculocutaneous telangiectasia and frequent pulmonary infection. Pediatrics 21: 526–554

Boder E, Sedgwick RP (1963) Ataxia-telangiectasia: a review of 101 cases. Proc Third Int Study Group, Child Neurol, Cerebral Palsy, Oxford 1962. Heinemann Medical Books, London

Evrard P, Beya A, Provis M (1990) Denise Louis-Bar. In: Ashwal S (ed) *The Founders of Child Neurology*. Norman Publishing, San Fransisco, pp. 774–777

Louis-Bar D (1941) Sur un syndrome progressif comprenant des télangiectasies capillaires cutanées et conjunctivales symétriques, à disposition naevoide et des troubles cérébelleux. Confin Neurol (Basel) 4: 32–42

Pelc S, Vis H (1960) Ataxie familiale avec télangiectasies oculaires (Syndrome de D Louis-Bar). Acta Neur Psych Belg 60: 905–922

Syllada L, Henner K (1926) Contribution à l'étude de l'indépendance de l'athétose double idiopathique et congénitale. Atteinte familiale, syndrome dystrophique, signe du réseau vasculaire conjonctival, intégrité psychique. Rev Neur 1: 541–560

# MAFFUCCI, Angelo Maria

*(1847–1903)*

*Courtesy*: Nighat Ispahany, Librarian, The New York Academy of
Medicine

M AFFUCCI syndrome comprises multiple cutaneous haemangiomata in association with irregular cartilaginous fragments (encondromatoses) in the metaphyses and shafts of the long bones. The skin and bony lesions are asymmetrical and do not coincide anatomically; and the genetic basis, if any, is uncertain.

# Biography

M AFFUCCI was a distinguished Italian pathologist who undertook important investigations on tuberculosis during the second half of the nineteenth century.

Angelo Maria Maffucci was born into a farming family at Calitri near Avellino, Italy, on 27 October 1847. He was an energetic student, driven by curiosity, and graduated in medicine at Naples in 1872. He spent several years as a surgeon at the Hospital for Incurables and in 1882 became head of general pathology at Messina University. In the following year he was called to the chair of anatomical pathology at Catania and in 1884 became professorial head of the same speciality at the University of Pisa. He remained at Pisa until his retirement, achieving the status of president of the faculty of medicine.

Maffucci was a diligent researcher and a massive collection of his meticulous notes, which was found in his home after his death, now reposes in the Institute for the History of Medicine in Rome. He recorded details of classical chick experiments, in which he determined that avian tuberculosis had a different aetiology from the bovine and human forms. He also recognized that chick embryos had defence mechanisms which could destroy bacteria.

Maffucci was proud, sincere and genial and was widely respected by his colleagues. Nevertheless, his important scientific contributions attracted little attention outside Italy. He died in Pisa on 24 November 1903 at the age of 56 years.

# Nomenclature

I N 1881, in the Italian literature, Maffucci described a patient with "dyschondroplasia and multiple cutaneous haemangiomata". In 1889, possibly unaware of Maffucci's article, Kast and von Recklinghausen reported another example of the same disorder. Thereafter the condition was known by a variety of descriptive titles and by the conjoined eponym, "Kast-Maffucci". However, Kast's name was soon abandoned and the term "Maffucci syndrome" has gained acceptance. About 100 cases have been reported, including a series of 62 patients reviewed by Anderson (1965).

The centenary of Maffucci's first case description was commemorated in 1981 by Tilsley and Burden in a paper in which Maffucci's original captions were used in an account of a contemporary patient.

# References


Acocella G (1954) Il pensiero medico di Angelo Maffucci. 14th Congresso Internazionale di Storia della Medicina. Roma 13–20 September. pp. 129–138

Anderson IF (1965) Maffucci's syndrome: Report of a case with a review of the literature. S Afr Med J 39: 1066–1070

Costa A (1965) Ricardo di Angelo Maffucci. Archivio de Vecchi per l'anatomia patologica e le medicina clinica 45: 1–17

Maffucci A (1881) Di un caso di encondroma ed angioma multiplo. Contribuzione alla genesi embrionale dei tumori. Il Movimento Med Chir 3: 399–412

Tilsley DA, Burden PW (1981) A case of Maffucci's syndrome. Br J Dermatol 105: 331–336.

# MARCHESANI, Oswald

*(1900–1952)*

*From*: Münchener Medizinische Wochenschrift (1952) 94: 1131

*Courtesy*: MMV Medizin Verlag München

W EILL-MARCHESANI syndrome, or spherophakia-brachymorphia (*see* Weill, p. 189).

# Biography

M ARCHESANI was a prominent Austrian ophthalmologist who published extensively prior to World War II.

Oswald Marchesani was born on 1 May 1900 in Schwaz, Tirol, where his father was an attorney. He spent his boyhood in Bozen and studied medicine at Innsbruck and Freiburg. After obtaining his doctorate at Innsbruck in 1923 Marchesani was appointed as an assistant in the university eye clinic in that city. He moved to Munich in 1927 and commenced a period of great academic productivity; his work received recognition and in 1936 he was called to the chair of ophthalmology in Münster. After the armistice in 1945 he was appointed to the chair in Hamburg, which he occupied until his death in 1952.

Marchesani was especially interested in the anatomy and histology of developmental defects of the eye. He undertook studies of the morphology of glial tissue of the optic nerve and, for many years, was occupied with the problem of sympathetic ophthalmia. In the later stages of his career, Marchesani investigated ocular disorders which were associated with disease of the salivary glands and joints. His findings were often at variance with popular theory, but he was frequently proved to be correct in his assumptions and conclusions.

Outside his professional life Marchesani was an accomplished Alpinist and spent his spare time in the mountains, enjoying his favourite hobbies of skiing and climbing. He was described as a distinguished, compassionate physician and a splendid, upstanding human being. Marchesani died in Kiel in 1952 from periarteritis nodosa, which he himself diagnosed, at the early age of 51 years.

# Nomenclature

I N 1939 Marchesani documented a boy aged 8 years and 3 siblings in another family, all of whom had ectopia lentis. He recognised that although ectopia lentis was a feature of the Marfan syndrome, the patients whom he had studied had short fingers rather than arachnodactyly. He emphasised this point by including comparative photographs in his article, in which he termed the disorder, "brachydactyly and congenital spherophakia".

Prior to Marchesani's report, Weill (1932) had published a report of 8 patients with dislocated lenses, including a woman with stunted stature and short, rigid digits. Although he apparently did not realise that this patient had a distinctive disorder, Weill's priority was eventually recognised by the addition of his name to that of Marchesani in the title of the condition.

In his original article, Marchesani had commented on parental consanguinity in one of the families which he had reported, and he had also noted that the parents had short stature. It therefore seemed likely that the condition was inherited as an autosomal recessive trait with minor manifestations in obligatory heterozygotes. In 1955, Kloepfer and Rosenthal published an account of a large affected kindred in Louisiana and confirmed that obligate heterozygotes had stunted stature.

The Weill-Marchesani syndrome is comparatively common in the Amish religious isolate of the USA, and McGavic (1966) has suggested the overall short stature of this group might be a reflection of the large number of heterozygotes in this community.

# References

Kloepfer HW, Rosenthal SW (1955) Possible genetic Carriers in the spherophakia-brachymorphia syndrome. Am J Hum Genet 7: 398–425

Marchesani O (1939) Brachydaktylie und angeborene Kugellinse als Systemerkrankung. Klin Monatsbl Augenheilkd 103: 392–406

McGavic JS (1966) Weill-Marchesani syndrome: brachymorphism and ectopia lentis. Am J Ophthamol 62: 820–823

Obituary (1952) Klin Monatsbl Augenheilkd 120: 653–654

Obituary (1952) Munch Med Woch 94: 1130–1131

Weill G (1932) Ectopie du cristallins et malformations générales. Ann Ocul 169: 21–44

# MARTIN, James Purdon

## *(1893–1984)*

*Courtesy*: G. Davenport, Librarian, Royal College of Physicians,
London

**M**ARTIN-BELL or Fragile X syndrome is a common form of X-linked mental retardation in which characteristic chromosomal fragile sites can be demonstrated.

# Biography

**M**ARTIN was a senior British neurologist during the middle period of the twentieth century.

James Purdon Martin was born on 11 June 1893 on a farm near Jordanstown, County Antrim, Ulster, Ireland. In 1912, after completing his schooling at the Belfast Royal Academical Institution, he took up a scholarship in modern languages at Queen's University, Belfast. A change of academic direction lead to Martin's graduation in 1915 with a first class degree in mathematics. After rejection by the armed forces because of chronic psoriasis, he returned to his university and embarked upon medical studies.

Martin qualified in medicine in 1920, after 8 happy years as an undergraduate, and undertook an internship in Liverpool. Many of his patients had the post encephalitic sequelae of the influenza epidemic and the experience gained in this field initiated his life-long interest in the pathophysiology of the basal ganglia of the brain. Armed with this special expertise, Martin moved to London for higher training and in 1925 he was appointed as consultant at the National Hospital for Nervous Diseases, Queen Square. He spent his whole career at this hospital, becoming Dean of the Medical School in 1944 and holding consultant posts at the Hammersmith, Bolingbroke and Whipps Cross Hospitals. During World War II, Martin served as neurologist to Eastern Command. After his retirement from the National Hospital in 1959, he spent a year as visiting professor at the University of Colorado, Denver, USA.

The monograph *The Basal Ganglia and Posture*, which was published in 1967, encapsulated Martin's research interests. This work was based upon his extensive experience with patients at the Highlands Hospital, Winchmore Hill, London, who were afflicted with Parkinsonism consequent upon encephalitis lethargica. Martin also edited the 8th and 9th editions of *Price's Textbook of Medicine*.

Martin retained his Ulster accent throughout his life. Although this attribute added to his charm he was a somewhat forbidding figure, held in awe by his juniors. Nevertheless, he was respected for his high intelligence, wide knowledge and great clinical prowess.

During his residency at the National Hospital, Martin married his colleague Marjorie Blandy. She was the first woman registrar at this hospital and as her career was placed in jeopardy by this contravention of contemporary values, they married in secret. She died in 1937 and a decade later he married Janet Ferguson, whom he had known while studying mathematics as an undergraduate in Belfast.

In 1983, at the age of 89 years, Martin was honoured by the award of a Doctorate of Science from Queen's University, Belfast. He died in the National Hospital in the following year, at the age of 90 years.

# Nomenclature

**I**N accordance with the fashion of the time, Martin preferred to use his second forename as a component of the non-hyphenated surname Purdon Martin and this format was generally favoured in his interaction with his friends and colleagues. Nevertheless, in his publications and in the title of the condition which bears his name, the simple form Martin is employed.

In 1943, in collaboration with Julia Bell of the Galton laboratory, London, Martin published details of a family in which several males had inherited mental retardation as an X-linked trait. Apart from delineating the condition, their recognition of X-linkage also served to explain the excess of males which had been observed in many institutions for the mentally retarded.

Additional reports of X-linked mental retardation followed, including a review from Canada published by Renpenning and his colleagues in 1962. Thereafter, the triple eponym enjoyed some popularity, but with the discovery of the fragile site on the X chromosome and the recognition of genetic heterogeneity, the nosological situation became very confused. At an International Congress on X-linked mental retardation, reported in 1984, it was agreed that the conjoined eponym "Martin-Bell" would apply to the form of the disorder with the fragile X chromosome, while the name "Renpenning" would be retained for X-linked mental retardation which was either undifferentiated or associated with stunted stature and macrocephaly. Despite this clarification there is still terminological confusion in this field (*see The Man Behind the Syndrome* Bell, p. 15, Renpenning, p. 224).

# References

Martin JP, Bell J (1943) A pedigree of mental defect showing sex linkage. J Neurol Psychiatr 6: 154–157

Obituary (1984) James Purdon Martin. Lancet 1: 1135–1136

Obituary (1984) JP Martin. Br Med J 288: 1698

Opitz JM, Sutherland GR (1984) Conference Report: International workshop on the fragile X and X-linked mental retardation. Am J Med Genet 17: 5–94

Renpenning H, Gerrard JW, Zaleski WA, Tabata T (1962) Familial sex-linked mental retardation. Can Med Assoc J 87: 954–956

# NETTLESHIP, Edward
## (1845–1913)

N ETTLESHIP syndrome, or urticaria pigmentosa, is a chronic familial skin disorder, which is characterised by widespread pigmented macules and nodules.

Nettleship-Falls syndrome, or ocular albinism type I (*see* Falls, p. 57).

# Biography

N ETTLESHIP was an academic ophthalmologist in London at the end of the nineteenth century. He is regarded as the founder of British ophthalmological genetics.

Edward Nettleship was born in Kettering, Northamptonshire, England, where his father was a solicitor. The influences which shaped his character are readily discernable in his family background. For instance, his parents and brothers had considerable talents and were involved with music, the arts and sciences, while his maternal grandfather was a clergyman and school headmaster. Nettleship's mother, an invalid, had great strength of character and strong views on the upbringing of her sons. She emphasised intellectual training and forbade outdoor games and frivolity. A relative described Nettleship's home as "an abode of plain living and high thinking, with all the essentials of cultured country life".

Nettleship developed an interest in natural history while attending Kettering Grammar School and spent his free time in woods and fields. He was attracted to ornithology and was nicknamed by his brothers "Bird-bearing Ned". He was destined to become a farmer (unlike his brothers, 2 of whom were Oxford dons) and he entered the Royal Agricultural College at Cirencester in 1862, moved to the Royal Veterinary College, Camden Town in 1863 and thence to King's College, London. In addition to his veterinary qualifications, Nettleship obtained medical status as a licentiate of the London Society of Apothecaries. He spent a few months as professor of Veterinary Surgery at the Cirencester College and then turned to medicine, proceeding to the Fellowship of the Royal College of Surgeons in 1870, while working under Jonathan Hutchinson at the London Hospital. Nettleship was interested in ophthalmology and he became a clinical assistant at Moorfields Eye Hospital, continuing his association with Hutchinson and becoming a close friend of Waren Tay (*see The Man Behind the Syndrome* p. 165). In 1873 Nettleship also became medical superintendent of the Ophthalmia Schools in Bow; here he used his influence to ameliorate the problems experienced by pauper children in London.

In 1878 Nettleship was appointed as ophthalmic surgeon at St Thomas's Hospital, where he remained for almost two decades. His department was held in high repute and his personal qualities received recognition when he was elected as dean of the medical school in 1888. He held concurrent appointments at Moorfields Eye Hospital until his retirement in 1898 and also had a highly successful private ophthalmological practice in Wimpole Street.

Nettleship was the author of the standard undergraduate textbook *The Student's Guide to Diseases of the Eye*, which went into 5 editions, and he also published numerous articles in the ophthalmological literature over a 40-year period. His review *On Some Hereditary Diseases of the Eye*, which was published in 1909, is a classical account of early concepts concerning ophthalmological genetics. In the same year he delineated ocular albinism, which he differentiated from conventional generalised albinism. This X-linked disorder is sometimes termed the "Nettleship-Falls syndrome" (*see* Falls, p. 57).

In addition to his reputation as a scientist, Nettleship was also widely regarded for his postgraduate teaching abilities.

He set high standards of clinical observation and documentation; he did not suffer fools gladly but his enthusiasm and personality attracted the best students.

Nettleship retired to his country home in Hindhead, Surrey, at the age of 57 years in order to devote himself to scientific study. He concentrated on inherited eye disease and during the next decade he accumulated a great quantity of pedigree data; much of this meticulously compiled material is now in the library of the Institute of Ophthalmology, London. His work was recognised by the award of the medal of the Ophthalmological Society in 1909 and in 1912 by his elevation to fellowship of the Royal Society. Nettleship died in Hindhead on 30 October 1913, at the age of 68 years.

# Nomenclature

I T is paradoxical that although Nettleship was a world figure in ophthalmology and a founder of ophthalmic genetics, his claim to eponymous immortality lies in the field of dermatological genetics. It is also of interest that Nettleship delineated the disorder which bears his name in a minor article written during the initial phase of his career. By contrast, his monumental contributions to the understanding of the genetic basis of conditions, such as albinism and retinitis pigmentosa, which he made more than 40 years later, now belong to the classical but archaic ophthalmological literature.

In 1869, while employed as assistant to Jonathan Hutchinson at the London Hospital, Nettleship published a brief report concerning a child with an unusual form of urticaria. He gave a graphic account of the skin lesions of Emile P, aged 2 years, living at Blackheath Hill, who was admitted to the hospital on 27 July 1879, with a chronic urticarial skin eruption, which had appeared when she was aged 3 months. Nettleship stated "A remarkable feature is that the wheals have brown stains. The appearance produced is very singular as the child, being of beautifully fair skin, the brown patches are the more conspicuous. At first sight the diagnosis of chloasma would occur to most, but the entire absence of branniness and the early age of the patient are fatal to this suspicion. It has also been set at rest by the microscope. The urticarious irritability of the child's skin is proved at once on scratching it." Nettleship noted "There was no history of urticaria in other members of the family".

Early authors employed the eponymous form "Nettleship syndrome" but the title "urticaria pigmentosa" subsequently came into use. The term "urticaria xanthelasmoidea" has also been used but this designation is now redundant. It is now recognised that the condition represents the skin manifestation of mastocytosis or mast cell disease. There are several reports which are consistent with autosomal dominant inheritance and concordant monozygotic twins have been documented. Non-penetrance has been reported and there is also some evidence for genetic heterogeneity.

# References


Fowler JR, Parsley WM, Cotter PG (1986) Familial urticaria pigmentosa. Arch Derm 122: 80–81

Jay B (1983) Remembrances of things past. Survey of Ophthalmology 27/4: 264–268

Lawford JB (1923) British Masters of Ophthalmology. Edward Nettleship. Br J Ophth 7: 1–9

Nettleship E (1869) Rare form of urticaria. Br Med J 2: 323–324

Nettleship E (1909) On some hereditary diseases of the eye. Trans Ophthalmol Soc, UK 29: 57–198

Selmanowitz VJ, Orentreich N (1970) Mastocytosis: a clinical genetic evaluation. J Hered 61: 91–94

# NOONAN, Jacqueline A
*(b. 1928)*

N OONAN syndrome comprises stunted stature, webbed neck, a characteristic facies and cardiac anomalies, notably pulmonary stenosis.

# Biography

N OONAN is a distinguished paediatric cardiologist, who has enjoyed a long and successful career at the University of Kentucky College of Medicine, USA.

Jacqueline A Noonan was born on 28 October 1928 in Burlington, Vermont, USA and spent her childhood in Hartford, Connecticut. She obtained a degree in chemistry from the Albertus Magnus College, New Haven in 1950 and qualified in medicine in 1954 at the University of Vermont.

Following an internship at the Memorial Hospital, University of North Carolina, Noonan completed paediatric residency training at the Children's Hospital, Cincinnati, Ohio. She then trained in paediatric cardiology at the Children's Hospital, Boston, under Dr Alexander Nadas, and in 1959 she was appointed as the first paediatric cardiologist at the University of Iowa. Noonan moved to the newly developed medical school at the University of Kentucky in 1961; in 1969 she was appointed as professor in her speciality and since 1974 she has also been chairman at the department of paediatrics.

Noonan has had an active academic career, publishing numerous articles, reviews and chapters in the field of cardiology. She has served on several professional boards and committees and acted as a reviewer for a number of specialist medical journals. She has been an invited speaker at many international meetings, and has received numerous honours and awards.

# Nomenclature

W HILE at Iowa, Noonan, together with Dorothy Ehmke, conducted a clinical investigation, which included 833 children with congenital heart disease who were carefully studied for the presence of additional extra-cardiac anomalies. Nine of these children had a similar characteristic facies and pulmonary stenosis. These patients were reported by Noonan to the Midwest Society of Paediatric Research held in Cincinnati in 1962.

During the course of Noonan's investigations in Iowa, John Opitz was a medical student and first year resident in the Department of Paediatrics. Subsequently, he went to Madison, Wisconsin, where he noted a number of other patients who were similar to those described by Noonan. These individuals had small stature, hypertelorism, mild mental retardation and often ptosis and skeletal abnormalities, as well as undescended testes in the males. Pulmonary stenosis was the most frequent cardiac defect. The chromosomes in these patients were normal and Opitz considered that they could clearly be distinguished from Ullrich-Turner syndrome. The designation "male Turner" and "female pseudo-Turner" syndromes had a transient vogue at this time. In 1985 Opitz proposed the name "Noonan syndrome" to denote such patients and presented a paper on this subject at the USA Society for Paediatric Research.

In 1968 the American Journal of Diseases of Childhood published a series of papers on the condition, including one by Noonan. She entitled her paper "Hypertelorism with Turner phenotype" but the editor used the term "Noonan syndrome" for a running title for 4 of the 5 papers in that issue. The eponym was firmly established in 1971 at the fourth conference on the clinical delineation of birth defects, Baltimore, where several papers on Noonan syndrome were presented.

An early description of the Noonan syndrome is credited to Kobilinsky (1883), a medical student at the Russian/ Estonian University of Dorpat; the original illustration of this patient was republished by Opitz and Pallister in 1979. The phenotype had also been documented by Weissenberg in 1928. In a historical review of the Noonan syndrome, Cole (1980) suggested that the blacksmith depicted in Albright's painting *Among Those Left* might have had the disorder.

# References


Cole RB (1980) Noonan's Syndrome: A Historical Perspective. Pediatrics 66: 468–469

Kobilinsky O (1883) Ueber eine flughautahnliche Ausbreitung am Halse. Arch Anthropol 14: 343

Mendez HMM, Opitz JM (1985) Noonan syndrome: a review. Am J Med Genet 21: 493–506

Noonan JA (1968) Hypertelorism with Turner phenotype. A new syndrome with associated congenital heart disease. Am J Dis Child 116: 373–380

Noonan JA, Ehmke DA (1963) Associated non-cardiac malformation in children with congenital heart disease. Abstract J Pediatr 63: 468–470

Opitz JM (1985) The Noonan syndrome (editorial). Am J Med Genet 21: 515–518

Opitz JM, Pallister PD (1979) Brief historical note: the concept of gonadal dysgenesis. Am J Med Genet 4: 333–343

Weissenberg S (1928) Eine eigentumliche Hautfaltenbildung am Halse. Anthropol Anz 5: 141–144

# OGUCHI, Chuta

*(1875–1945)*

*Courtesy*: Professor Y. Oguchi, Japan

O GUCHI disease is a form of congenital non-progressive night blindness which is associated with diffuse grey or yellow discoloration of the fundus. Inheritance is autosomal recessive.

# Biography

C HUTA OGUCHI was a distinguished Japanese ophthalmologist during the first half of the twentieth century.

Chuta Oguchi was born on 6 January 1875 in Ueda, Nagano prefecture, Japan. He attended the Ueda primary school, and moved to Tokyo where he entered the Saiseigakusha School. After graduation Oguchi passed the national medical board examination and trained in ophthalmology under Professor Testuzo Suda and subsequently with Professor Koumoto at the University of Tokyo. In 1895, Oguchi volunteered for military service; he was commissioned as a third class army doctor with the First Guards Infantry Regiment in Taiwan and was decorated with the Sixth Order of the Sacred Treasure for his contribution during the Sino-Japanese War.

Oguchi worked at the Nagano Public Hospital in 1896 and then at the Tokyo Army Hospital from December 1897. He was promoted to second class army doctor in 1899, transferred to Kokufudai Army Hospital, and then to Shonan Hospital in Yokosuka as chief of the department of ophthalmology. Oguchi became chief of the department of ophthalmology at the Tainan Government Hospital, Taiwan in 1903 and served in the Russo-Japanese War between 1904 and 1905.

In 1911 Oguchi became a faculty member in the army medical school, and he was then invited to take up the post of Professor of Ophthalmology at South Manchuria School of Medicine before spending sabbatical years in Germany at Heidelberg and Munich. During this period he published articles in the German language in which he documented the effects of ocular trauma sustained during the Russo-Japanese war. In 1914, after attending an international conference in London, Oguchi returned to Japan where he was awarded the degree of PhD from the University of Tokyo.

In 1919 Oguchi was appointed as associate professor at the Aichi Prefectural Medical School and Chief of the department of ophthalmology at the Prefectural Hospital. He received professorial status in 1922, when the school was upgraded, and he became Dean in 1926. In 1931 Oguchi became professor at Nagoyi Medical College and he received an award in 1933 for the discovery of the condition which bears his name. His other major academic contributions were the construction of the first colour perception testing chart in Japan and the system which became the standard for visual acuity tests. Oguchi was appointed as professor at Nagoya Imperial University in 1939 and after retiring in the same year, he continued teaching at the university with honorary professorial status.

Election as a member of the International Ophthalmological Society Board of Directors took place in 1929. Oguchi occupied this position until 1937, after which he became a member of the International Society of Prevention of Trachoma. In the later stages of his career, Oguchi served as trustee and superintendent of the Japanese Ophthalmological Society, President of the Society of Japanese Ophthalmologists, and President of the Japanese Society of Genetics. In 1936 he presented a review of inherited ocular disorders at the annual meeting of the Japanese Society of Genetics and in 1936, a review on ocular trauma at the 40th annual meeting of the Japanese Ophthalmological Society. Oguchi also wrote a book on the history of ophthalmology in Japan, for which he was decorated with the order of the Sacred Treasure, second class.

Oguchi died on 23 July 1945 at the age of 70 years. Several of his descendents pursued careers in medicine, including his eldest son, Tadao, who was chief of Ophthalmology at Ookubo Hospital, and his grandson, Nobuo, presently director of the Oguchi eye clinic in Seto city. His son-in-law, Takehisa, was the former Director of the Kanagawa Ophthalmological Society and his sons (Chuta's grandsons) are Yoshihisa, the Professor of Ophthalmology at Keio University and Kazuhisa, the director of the Oguchi eye clinic in Kawasaki city.

# Nomenclature

I N 1907 and 1910 Oguchi was the author of articles in the Japanese language, in which he gave accounts of the syndrome of night blindness and fundal changes which now bears his name. He published a further article in the German literature in 1912 in which he used the term "hemeralopie", in order to emphasise the fact that vision was defective in the twilight. Many affected persons were identified in Japan and by 1924, Takagi and Kawakami were able to analyse 56 cases, adding three of their own and commenting on the familial nature of the disorder.

Scheerer documented the first non-Japanese cases in 1927 and thereafter details of others were published. The eponym "Oguchi" was introduced in the north American literature by Klein in 1939 and the Latin descriptive term "fundus albipunctatus cum hemeralopie" was used by Franceschetti and Chome-Bercioux in 1951. The eponymic format was preferred by Francois and colleagues in 1956 in an article in the French language in which they reviewed 72 cases from Japan plus 23 from other parts of the world. Academic interest is now focussed on the elucidation of the underlying gene defect and Oguchi's name has appeared on the majority of articles which have addressed this problem. Molecular investigations in an affected Indian kindred have provided evidence that the determinant gene is situated on the distal region of the long arm of chromosome 2 (Maw et al. 1995).

Oguchi syndrome is sometimes confused with the condition known as congenital stationary night blindness or hemeralopia. The most obvious difference is the autosomal dominant mode of transmission in this latter disorder, which has been ascribed in some families to a faulty gene on the short arm of chromosome 4.

# References


Bordley J (1908) A family of hemeralopes. Bull Johns Hopkins Hosp 19: 278–281

Carr RE, Gouras P (1965) Oguchi's disease. Arch Ophthalmol 73: 646–656

Franceschetti A, Chome-Bercioux N (1951) Fundus albipunctatus cum hemeralopie (cas stationnaire depuis 49 ans). Ophthalmologica 121: 185–193

Francois J, Verriest G, De Rouck A (1956) La maladie d'Oguchi. Ophthalmologica 131: 1–40

Klein BA (1939) A case of so-called Oguchi's disease in the USA. Am J Ophthalmol 22: 953–955

Maw MA, John S, Jablonka S, Müller B, Kumaramanickavel, Oehlmann R, Denton MJ, Gal A (1995) Oguchi disease: suggestion of linkage to markers on chromosome 2q. J Med Genet 32: 396–398

Oguchi C (1907) One kind of night blindness. Acta Soc ophthalmol Jap 11: 123–134

Oguchi C (1910) Another case of one kind of night blindness. Acta Soc ophthalmol Jap 14: 636–639

Oguchi C (1912) Über die eigenartige Hemeralopie mit diffuser weissgräulicher Verfärbung des Augenhintergrundes. Graefes Arch Ophthmol 81: 109–117

Scheerer R (1927) Der erste sichere Fall von Oguchischer Krankheit mit Mizuoschem Phänomen ausserhalb Japans. Klin Mbl Augenheilk 78: 811–813

Taragi R. Kawakami R (1924) Über das Wesen der Oguchischen Krankheit. Klin Mbl Augenheilk 72: 349–371

# OLLENDORFF, Helene

## (b. 1899)

B USCHKE-OLLENDORFF syndrome or dermatofibrosis lenticularis disseminata with osteopoikilosis
is an uncommon autosomal dominant disorder in
which indurated patchy skin lesions are associated with multiple sclerotic foci in the skeleton.

# Biography

O LLENDORFF, using her married name Curth, had a
distinguished career in dermatology in the USA
during the middle part of the twentieth century.

Helene Ollendorff was born on 28 February 1899, in Breslau,
Germany. Her father was a prominent member of the legal
profession and her mother was a well-known feminist leader.
Ollendorff studied medicine at the Universities of Freiburg,
Munich and Breslau and after qualification in 1923 trained in
dermatology under Professor A. Buschke at the Rudolf
Virchow Hospital in Berlin (*see* Buschke, p. 37). She became
Buschke's assistant and subsequently married his other as-
sistant, Dr. William Curth.

In 1931 the Curths moved to the USA with their baby
daughter. They commenced private practice in New York and
concurrently continued with their careers in dermatology at
Columbia University. After her marriage, Ollendorff took her
husband's surname, retaining her own as an unhyphenated
forename and her publications in the field of dermatology
appeared under the name H.O. Curth. She gained international recognition for her work on acanthosis nigricans and
travelled widely in connection with investigations in this
field. At the time of her retirement Dr U.W. Schnyder of
Zurich said of her that "her critical sense for the essential
point as well as her feminine warmth are characteristics of
this successful dermatologist".

# Nomenclature

I N 1928 Buschke and Ollendorff documented a
woman, aged 41 years, with pea-sized papules over
the posterior aspect of the trunk, thighs and upper
arms. Radiologically, multiple discrete sclerotic foci were
evident in the pelvis, shoulder girdle and ends of the long
bones. These skeletal abnormalities, which were innocuous
and clinically silent, had initially been documented by Stieda
in 1905, while the term "ostéopoecilie" (anglicised as osteopoikilosis) was coined by Ledoux-Lebard et al. in 1916.
The current term for the condition "dermatofibrosis lenticularis disseminata with osteopoikilosis" was employed by
Ollendorff in 1934, under her married name Curth. Further
reports followed and the condition is now widely recognised.
The conjoined eponym is well known, although the descriptive title is generally preferred.

# References


Buschke A, Ollendorff H (1928) Ein Fall von Dermatofibrosis lenticularis dis-
seminata und Osteopathia condensans disseminata. Derm Wschr 86:
257–262
Curth HO (1934) Dermatofibrosis lenticularis disseminata and osteopoikilosis.
Arch Derm Syph 30: 552–560
Ledoux-Lebard R, Chabeneix, Dessane (1916) L'ostéopoecilie: Forme nouvelle
d'ostéite condensante généralisée sans symptomes cliniques. J Radiol
Electrol 2: 133–134
Schnyder UW (1964) Zum 65. Geburtstage von Helen Ollendorff Curth.
Hautarzt 15: 96–97
Stieda A (1905) Ueber umschriebene Knochenverdichtungen im Bereich der
Substantia spongiosa im Roentgenbilde. Beitr Klin Chir 45: 700–703

# PARKINSON, Sir John
## *(1885–1976)*

*Courtesy*: G. Davenport, Librarian, Royal College of Physicians, London

# W

OLFF-PARKINSON-WHITE syndrome (*see* White, p. 193)

# Biography

PARKINSON was a distinguished cardiologist on the staff of the London Hospital and the National Heart Hospital during the twentieth century. He was knighted for his services to his speciality.

[1]John Parkinson was born on 10 February 1885 in Thornton-le-Fylde, a village close to the northwestern coast of Lancashire, England. His family were well established in that locality, and his father served as a magistrate. Parkinson attended Manchester Grammar School, an institution famous at the time for rigorous academic standards, and then proceeded to University College, London. He subsequently trained in medicine at the University of Freiburg and the London Hospital, qualifying in 1907, with several prizes, and obtaining a doctorate in 1910.

Parkinson spent the early part of his career at the London Hospital, where he was assistant to Sir James Mackenzie in the department of cardiology. In 1914, when World War I broke out, Parkinson joined the Royal Army Medical Corps and served in casualty clearing stations on the battlefields of France. By 1917 he had achieved the rank of major, and he commanded a military heart centre in Rouen. Following demobilisation in 1919, Parkinson returned to the London Hospital, where he took charge of the cardiac department and eventually achieved consultant status. He was also appointed to the consulting staff of the National Heart Hospital, London and as civilian cardiologist to the Royal Air Force for the period 1931 to 1956.

His researches were clinically orientated and mainly aimed at improving methods of diagnosis of heart disease. Parkinson was a proponent of the introduction of modern diagnostic techniques, notably electrocardiography and the radiological examination of the heart and he concerned himself with dysrhythmias, rheumatic fever, the cardiac effects of thyroid disturbance and coronary artery disease. He also supported the advances in cardiac surgery which took place towards the end of his career. In addition to diagnosis, Parkinson was involved in the development of therapeutic techniques, including the use of strychnine in heart failure and adrenaline in Stokes-Adams attacks. In 1917, during his military service, he demonstrated that digitalis therapy had no value in the management of "soldier's heart" and he emphasised that this malady had its origins in the nervous system. Parkinson also confirmed the beneficial activity of quinidine in certain arrhythmias and of digitalis in all forms of cardiac failure. He introduced mercury suppositories as a diuretic and advocated adequate rest and analgesia in heart failure.

Parkinson's clinical acumen, publications and teaching skills attracted international recognition and he became president of the European Society of Cardiology, an honorary member of the International Congress of Cardiology and an honorary fellow of the American College of Physicians. He received the golden stethoscope of the International Cardiology Foundation, the Fothergill gold medal of the Medical Society of London, and the Moxon medal of the Royal College of Physicians of London. Parkinson's life-long aim was the firm establishment of cardiology as an acknowledged speciality within medicine; he achieved his goal and in 1948 he was knighted by King George VI for his services.

Concerning his character and attributes, Parkinson's biographer wrote "By nature, John Parkinson was a shy man. He enjoyed the company of close friends, when he was a generous host, but unless obligatory, he preferred not to dwell long in the company of strangers, other than those with a common interest, and especially the young seeking his advice; to them he would devote unlimited time and help. Those whom he had befriended and trained carried a passport to the world's foremost cardiological centres."

Another biographer commented "Intolerant of pomposity, he was wholly unspoiled by the many honours bestowed on him, and though he had under his care many of the world's greatest men he had just as much time for the poor and needy, and his invariable kindness and cheerfulness at the bedside brought great comfort to countless sick folk."

Parkinson married Clara, daughter of Alfred Le Brocq of St Helier, in 1917. The couple had 4 daughters and a son, who was killed in action while serving in the RAF in 1942. Parkinson retired from practice in 1960, when he was 74 years of age and lived in a flat near Baker Street, London. He died in 1976 at the age of 91 years.

# Nomenclature

IN 1930 John Parkinson collaborated with his North American colleagues, Louis Wolff and Paul White, in the description of the cardiac conduction defect which bears their triple eponym. At this time White was the head of the cardiology department at Massachusetts General Hospital, Boston, while Wolff was the chief of the electrocardiographic service at the same hospital.

The original article by Wolff, Parkinson and White contained an account of a form of bundle branch block in 11 healthy young adults who were subject to episodes of paroxysmal tachycardia. The relative contributions of the authors is uncertain; one patient had been seen at the London Hospital, while the majority of the others were examined in Boston. In their article Wolff, Parkinson and White alluded to a previous report of an affected student, aged 19 years, published by Wedd in 1921 and a similar example, complicated by mitral stenosis, described by Wilson in 1915. Although attention was drawn to the general absence of any underlying cardiac abnormality, there was no mention of the possible familial nature of the disorder.

The Wolff-Parkinson-White syndrome (WPW) achieved early and wide recognition and the eponymic title is still retained (*see* White, p. 193).

# References


Obituary (1976) John Parkinson. Lancet 1: 1359
Obituary (1976). Br Med J 1: 1536
Wedd AM (1921) Paroxysmal tachycardia. Arch Int Med 27: 571–590
Wilson FN (1915) A case in which the vagus influenced the form of the ventricular complex of the electrocardiogram. Arch Int Med 16: 1008–1027
Wolff L, Parkinson J, White PD (1930) Bundle-branch block with short PR interval in healthy young people prone to paroxysmal tachycardia. Am Heart J 5/6: 685–704


---

[1]Sir John Parkinson (1885–1976) is sometimes confused with James Parkinson (1755–1824) whose eponym is attached to the neurological disorder which is also known as paralysis agitans. Their respective families lived in different parts of England and they were not directly related to each other.

# PARRY, Caleb Hillier

*(1755–1822)*

PARRY-ROMBERG syndrome (*see* Romberg, p. 147).

# Biography

P**ARRY was a physician in Bath, England, at the turn of the eighteenth century. He is remembered for his accurate explanation of the pathophysiology of ischaemic heart disease.**

Caleb Hillier Parry was born on 21 October 1755 in Cirencester, Gloucestershire, England, where his father was a non-comformist minister. He was the eldest of 10 children, and together with several of his siblings, he was educated at Cirencester Grammar School. Edward Jenner, who developed vaccination for the prevention of smallpox, was a contemporary at Parry's school and the two became close friends. Many years later, in 1798, Jenner dedicated his seminal treatise *Inquiry into the Causes and Effects of Variolae Vaccinae* to Parry, and Parry reciprocated in 1814 when he dedicated his own publication on rabies to his friend Jenner.

At the age of 15 years, Parry moved to the Dissenters' Academy, Warrington, Lancashire and in 1773 he became a medical student at the University of Edinburgh. After spending 2 years in London with Dr Thomas Denham of the Middlesex Hospital, Parry returned to Edinburgh, obtaining his medical doctorate in 1778 with a thesis on the subject of rabies. Parry married in 1778 and in the following year he commenced general practice in Bath. He was also appointed as physician to the Puerperal Charity Hospital and subsequently to the Casualty Hospital in that city.

Parry was interested in experimental physiology and he conducted many investigations on the circulation and the effects of impairment of the vascular supply. His most notable contribution was a treatise published in 1799 entitled *An Inquiry into the Symptoms and Causes of Syncope Anginosa, commonly called Angina Pectoris; illustrated by Dissections.* In his work, which was based in part upon lectures which he had given to the Gloucestershire Medical Society, he expounded the concept that ischaemic heart disease resulted from energy demands of the myocardium, which the vascular system was unable to supply. Other publications followed, culminating in a review in 1816 of the results of his animal experiments into the nature of the arterial pulse, in which he correctly concluded that the pulse wave was generated by the rhythmical contractions of the left ventricle. Parry also made the first description of thyrotoxic exophthalmus, predating the classical accounts which were subsequently published by Graves and von Basedow.

Parry had wide interests and an active mind, and he spent much of his spare time collecting fossils. The conclusions which he drew from his collection were published in 1781 as *Proposals for a History of Fossils in Gloucestershire.* He was also active in farming and was honoured for his promotion of the wool industry. His broad scientific contributions were recognised by his election to the Royal Society.

As a successful physician he had distinguished patients including the astronomer Herschel (1738–1822), and Admiral Rodney of the Royal Navy (1718–1792). Parry also enjoyed a happy marriage and family life. One of his sons, William Edward, became an Arctic explorer and naval admiral, while another, Charles Henry, became physician to the Royal United Hospital, Bath.

Parry suffered a stroke in 1816 which left him with aphasia and right hemiplegia, and he died on 9 March 1822 at his home in Sion Place, Bath.

# Nomenclature

AFTER his stroke in 1816 Parry was unable to continue with his medical practice. Despite his disability, his intellect was unimpaired, and with the assistance of his dutiful daughter, he accumulated a collection of his writings and observations. These included an account of the condition now known as progressive hemifacial atrophy, to which his name is attached, together with that of Romberg. This description appeared in *Collections from Unpublished Papers* in 1825, 3 years after his death, published by his eldest son, the physician, Charles Henry Parry.

The disorder was again described in 1846 by Romberg, and the descriptive title "progressive hemifacial atrophy" was coined in 1871 by Eulenberg. Many reports followed, including a definitive review in 1983 by Lewkonia and Lowry. The aetiology of Parry-Romberg progressive hemifacial atrophy is still uncertain, and although autosomal dominant inheritance is possible in some families, the majority of affected persons are sporadic.

(For further details of the nomenclature, *see* Romberg, p. 147.)

# References

Eulenberg A (1871) *Lehrbuch der Functionellen Nervenkrankheiten auf Physiologischer Basis.* A Hirschwald, Berlin, p. 712

Kligfield P (1982) The early pathophysiologic understanding of angina pectoris. Am J Cardiol 50: 1433–1434

Lewkonia RM, Lowry RB (1983) Progressive hemifacial atrophy (Parry-Romberg syndrome) report with review of genetics and nosology. Am J Med Genet 14: 385–390

Parry CH (1825) *Collections from Unpublished Papers.* Underwood, London

Rolleston H (1925) Caleb Hillier Parry, MD, FRS. Ann M Hist 7: 205–215

Romberg MH (1846) *Trophoneurosen, Klinische Ergebnisse.* A. Förstner, Berlin

# PELGER, Karel

*(1885–1931)*

From: Nederlands Tijdschrift voor Geneeskunde (1931) 75: 4107

Courtesy: Nederlands Tijdschrift voor Geneeskunde, Amsterdam

**P**ELGER-HUËT anomaly is an innocuous autosomal dominant condition in which white blood cells have hyposegmented, misshapen nuclei.

# Biography

**P**ELGER was a general medical practitioner and a founder of haematology in Holland.

Karel Pelger was born on 14 April 1885 at Nijeveen, near Meppel, Holland. He qualified in medicine at the University of Groningen in 1912 and entered general practice in Amsterdam.

Pelger had an enquiring mind and a love of science and in his spare time he undertook microscopical studies of blood disorders. His enthusiasm led him to train in haematological techniques with Schilling in Berlin and his researches were facilitated by links with the faculty of medicine at the University of Amsterdam, where he was permitted to examine patients with blood disorders. Pelger emerged as the expert in his country in the field of morphological abnormalities of blood cells, and he was called upon to lecture widely on this topic. In this way he was able to demonstrate the importance of haematological investigation to his medical colleagues, and to bring haematology into general medicine.

He was a likeable, friendly person with a happy family life and despite his busy general practice and his haematological activities, Pelger put a great deal of effort in a farm which he owned at Havelte, Overijssel. He preferred life in the countryside but remained in Amsterdam because of his scientific interests.

Pelger planned to encapsulate his knowledge of clinical haematology in a textbook but in 1931, before this work could be completed, he died suddenly at his farm, aged 47 years.

# Nomenclature

**O**N 20 November 1927, at a paediatric meeting, Pelger demonstrated his haematological findings in 2 patients who had died with splenic enlargement and tuberculosis. He drew attention to abnormalities of the nuclei of their white blood cells and suggested that these changes were a component of the patient's condition. His findings were published in the Dutch literature in the following year. Thereafter, Huët of Hoog-Blaricum recognised the abnormality in several members of a family and suggested that the white cell changes were an innocuous hereditary trait. Shortly before Pelger's death, the familial white cell condition was brought to the attention of the Dutch Royal Academy of Science by Professor Hagmans and in 1932, after Pelger's death, Huët published an account of the disorder, in which he emphasised its heritability. The condition became widely known and the conjoined eponym was generally accepted.

The familial nature of the Pelger-Huët anomaly was emphasised in 1962, when Skendzel and Hoffman reported 41 affected persons in 7 kindreds. In the same year Ludden and Harvey drew attention to the presence of the Pelger-Huët anomaly in persons of Dutch descent living in Spokane, USA. The condition is not confined to persons of Dutch stock; for instance a large affected French-Canadian kindred living in Quebec was reported by Rioux et al. in 1968.

The Pelger-Huët anomaly occurs in rabbits, and breeding experiments have shown that the homozygote has skeletal abnormalities. Fortunately, the human homozygote is not similarly affected.

# References


Huët GJ (1932) Over een familiaire anomalie der leucocyten. Mschr Kindergeneesk 1: 173–181

Ludden TE, Harvey M (1962) Pelger-Huët anomaly of leukocytes: report of a case and survey of incidence. Am J Clin Path 37: 302–304

Obituary (1931): Ned Tijdschr Geneeskd 75/III: 4106–4108

Pelger K (1928) Demonstratie van een paar zeldzaam vorkomende typen van bloedlichampies en besprecking der patiënten. Ned Tijdschr Geneeskd 72: 1178–1182

Rioux E, St Arneault G, Brosseau C (1968) The Pelger-Huët anomaly of leukocytes: description of a Quebec kindred. Canad Med Assoc J 99: 621–624

Skendzel LP, Hoffman GC (1962) The Pelger anomaly of leukocytes: forty-one cases in seven families. Am J Clin Path 37: 294–301

# PENDRED, Vaughan

## *(1869–1946)*

Courtesy: Air Marshall Sir Lawrence Pendred, Leamington Spa,
England

**P**ENDRED syndrome is an autosomal recessive disorder comprising perceptive deafness and enlargement of the thyroid gland.

# Biography

**P**ENDRED trained in surgery but spent the greater part of his career as a general practitioner in Surrey, England.

Vaughan Pendred was born in Herne Hill, South East London, on 10 August 1869. His father, Vaughan Pendred, was a civil engineer and editor of the professional journal *The Engineer*. After attending Westminster School, Pendred studied medicine at the University of Newcastle-upon-Tyne and at Guy's Hospital, London. While still a student, he presented a case of arterio-venous aneurysm at the Physical Society. In 1894 he was appointed as house surgeon and obstetric resident at Guy's Hospital and he graduated with honours at the University of Durham in 1896.

In the early stages of his career, Pendred was interested in surgery and he obtained the Fellowship of the Royal College of Surgeons in 1896. He was then appointed as assistant in the otolaryngological department at Newcastle Infirmary. It was at this time that he reported a deaf-mute family with the condition which subsequently bore his name. Pendred received the doctorate of the University of Durham in 1901. Thereafter his career changed direction and after an appointment as assistant medical superintendent at the Fulham Hospital, London, he entered general practice in Buckingham. Pendred moved to Coventry and in 1913 to East Sheen, Surrey, where he spent the remainder of his working life.

Despite his obvious abilities Pendred did not have academic leanings and, apart from the report of the condition which bears his name, he wrote few, if any, medical articles. He took an active interest in professional matters, however, and he served on the Richmond division of the British Medical Association, and as honorary secretary and vice-president of the West London Medical Chirurgical Society. In Pendred's obituary, a member of the Society stated "He will be affectionately remembered for his lively contributions to post-war meetings of the Society when, with his rugged countenance he always had something fresh to say on the subject, or something challenging to provide in the discussions".

# Nomenclature

**A**T the end of the last century, while employed at the Newcastle-upon-Tyne Infirmary, Pendred documented two sisters with deaf-mutism and large goitres. His concise paper, which was published in 1896 in the *Lancet*, contained the following account of the disorder: "The curious association of deaf-mutism and goitre occurring in two members of a large family has induced me to record these cases. Why this association? Perhaps some readers of *The Lancet* may be able to throw some light on the cause of this combination of diseases: Absence of thyroid–cretinism; overgrowth of thyroid – deaf-mutism. I append the family history as recounted to me by the mother. The family is an Irish one, and the parents have been upwards of forty years resident in Durham. The father, aged sixty-six years, and the mother, aged sixty-seven years, are alive and healthy. They have had ten children, five sons and five daughters. In an epidemic of small-pox twenty-five year ago the whole family was attacked with the exception of the younger of the deaf-mutes, and four males and one female died, although all had been vaccinated, and that recently, as they were children. The remaining son and two of the daughters are healthy and vigorous. The first goitre case is the first-born of the family – a spare woman now aged thirty-eight years. She is deaf and can only mumble indistinctly; little care has been taken to educate her and so she is imbecile. The goitre is a large multi-lobular hard tumour, the greater part on the right side of the neck; from time to time she suffers from dyspnoeic attacks. The growth was first observed after the small-pox – i.e. at thirteen years of age. The second surviving girl is now aged twenty-eight years, and is the fifth of the family; she is a small, spare, intelligent woman, her expression being in marked contrast to her sister's. She is not absolutely deaf and can mumble incoherently; her education has been attended to with so much success that she has been "in service". The tumour is larger than in the other case, but is of the same character; it has been growing for about fifteen years, and during the last year has caused both dyspnoea and dysphagia, which have become so urgent that I have sent her today to Newcastle Infirmary for operation."

The association of deaf-mutism and goitre had previously been observed in large number of Swiss children by Bircher (1883) but it was recognised that a significant proportion of these persons were mute due to mental impairment rather than deafness. Early examples of the disorder which came to be known as the Pendred syndrome were documented by Purdon (1865) in Northern Ireland, and by Poncet (1865) in Mexico. The latter author was serving as a medical officer with the French Army during the reign of the Emperor Maximilian, and he encountered the condition in the endogamous inhabitants of the remote village of Noria, near Mazatlan in Sinaloa, Mexico.

After Pendred's classical account of the condition thirty years elapsed before the next report appeared in the medical literature. In 1927, while working in the medical unit of the London Hospital, Brain documented 5 sibships in the East End of London, containing 12 affected individuals, and suggested that inheritance was autosomal recessive. Almost 30 years later George Fraser restudied several of these persons as part of his monumental investigation into genetic deafness. In a paper in which he reported upon 207 affected families, he proposed the eponymous title "Pendred syndrome". This format is now firmly established.

# References

Bircher H (1883) *Der endemische Kropf und seine Beziehungen zur Taubstummheit und zum Cretinismus.* Schwabe, Basel

Brain WR (1927) Heredity in simple goitre. Quart J Med 20: 303–319

Fraser GR (1965) Association of congenital deafness with goitre (Pendred's syndrome). A study of 207 families. Ann Hum Genet Lond 28: 201–249

Obituary (1946) Br Med J 1: 259

Obituary (1946) West London Med J 51: 37

Pendred V (1896) Deaf-mutism and goitre. Lancet ii: 532

Poncet M (1865) Des mariages consanguins à la Noria (près Mazatlan), Sinaloa, Mexique. Recueil Mém Méd Chir Pharm Milit 14: 193

Purdon HS (1865) *Peculiarities of the Deaf and Dumb.* Greer, Belfast

# PERTHES, Georg Clemens

*(1869–1927)*

P
ERTHES disease, also known as Legg-Calvé-Perthes disease, is a form of osteochondritis of the femoral capital epiphysis which presents in childhood as hip pain.

# Biography

P
ERTHES was an orthopaedic surgeon and Professor of Surgery at the University of Tubingen, Germany in the early decades of the twentieth century.

Georg Clemens Perthes was born in 1869 in Moers, Rhineland. His mother died of tuberculosis while he was still young, and his father, a teacher, took the family to Davos in Switzerland, where he taught tubercular children. He then moved to Bonn, and after his untimely death, Georg Perthes was brought up by his maiden aunt, Agnes Perthes. Through family connections, Perthes met Trendelenburg, who influenced him in his choice of a career in medicine, and he qualified after studying at the universities of Freiburg, Bonn and Berlin.

In 1895 Perthes became an assistant to his mentor, Trendelenburg, in the city of Leipzig. He spent the period 1900–1901 in China as a surgeon in the German army and he took the opportunity to undertake radiological studies of the feet of Chinese women, which had been crushed and bound in the traditional manner. Perthes then returned to Leipzig where he made important surgical innovations, including the development of suction drainage for empyema and the use of a pneumatic cuff for haemostasis during limb operations. In the field of operative surgery, Perthes introduced new techniques for the stabilisation of recurrent dislocation of the shoulder joint and for the correction of limb malalignment. He also pioneered the use of x-rays in the treatment of warts, skin cancer and carcinoma of the breast.

Trendelenburg and Perthes had a warm relationship and held each other in mutual respect. The former had a happy outgoing personality while the latter had a quiet disposition. Nevertheless, despite his serious approach to life and a tendency to outspokenness in academic discussion, Perthes was liked and respected by his colleagues.

Perthes rejoined the German army in World War I, and after active service, moved to a base hospital where he treated many soldiers with wounds involving the peripheral nerves. His experience in this sphere was subsequently documented in texts which he wrote on the surgical management of injuries to the nerves and the mandible.

Towards the end of his career, Perthes was called to the chair of surgery at the University of Tubingen. In 1927, while enjoying a Christmas holiday in Arosa, Switzerland, Perthes died suddenly from a stroke at the age of 58 years.

# Nomenclature

T
HERE is a popular misconception that the name "Perthes" is the possessive form and an inverted comma is sometimes incorrectly inserted in the title of the disease process which bears this name. This problem has been compounded by the modern trend to avoid possessive eponyms and in the medical literature the erroneous format occasionally escapes editorial correction. In actual fact, the name "Perthes" is neither possessive nor plural.

In October 1910, while working with Trendelenburg at Leipzig, Perthes wrote an article on the subject of juvenile deforming arthritis. In the same year, by a singular concurrence, and unbeknown to one another, Jacques Calvé in France and Arthur Thornton Legg in the USA described the same disorder. Various combinations and orders of the triple eponym are sometimes used and the name of Waldenström is occasionally added. The descriptive title "osteochondritis deformans" has been employed for the disorder, but rightly or wrongly the simple form "Perthes disease" is generally preferred.

With the delineation of the phenotype and the accumulation of case reports, it has become apparent that there may be genetic factors in the pathogenesis of Perthes disease. Currently, the concept of multifactorial inheritance is favoured (*see* Calvé, p. 39).

Perthes' name is used in conjunction with that of Jüngling as an alternative title for osteitis cystica multiplex tuberculosa. This unusual slowly progressive inflammatory disorder of the digits and nose is now rarely encountered.

# References


Calvé J (1910) Sur une forme particulière de pseudo-coxalgie greffée sur des déformations caractéristiques de l'extremité supérieure du fémur. Rev Chir (Paris) 42: 54–84

Jüngling O (1919–21) Ostitis tuberculosa multiplex cystica (eine eigenartige Form der Knochentuberkulose). Fortschr Röntgen 27: 375–383

Legg AT (1910) An obscure affection of the hip-joint. Boston Med Surg J 162: 202–204

Perthes GC (1910) Ueber Arthritis deformans juvenilis. Dtsch Z Chir 107: 111–159

Perthes GC (1921) Über plastischen Daumenersatz insbesondere bei Verlust des ganzen Daumenstrahles. Arch Orthop Unfall-Chir 19: 198–214

Waldenström H (1920) Coxa plana. Osteochondritis deformans coxae, Calvé-Perthe'sche Krankheit, Legg's Disease. Z Chir 47: 539–542

# PEYRONIE, François de la

## *(1678–1747)*

*Courtesy*: G. Davenport, Librarian, Royal College of Physicians,
London

P EYRONIE disease is a male-limited autosomal dominant disorder in which curvature of the penis results from fibrous induration.

# Biography

P EYRONIE was the pre-eminent French surgeon of the eighteenth century. He served as Surgeon to King Louis XV and founded the Royal Academy of Surgeons of France.

François de la Peyronie was born on 15 January 1678 in Montpellier, France, where his father was a surgeon. He was educated at a Jesuit school and subsequently studied philosophy for 2 years, before training in surgery in his home town. He was a diligent student and qualified at the early age of 19 years, receiving congratulations from many of his fellow citizens of Montpellier.

Peyronie was sent to Paris to work under Mareschal, the chief of surgery at the Hôtel-Dieu, who was also surgeon to the King. He then returned to Montpellier, where he became an outstanding teacher of anatomy and surgery, receiving professorial status. During this period he served as surgeon-major in an army which was raised in order to put down a rebellion in Cévennes. Peyronie's reputation as a surgeon was established when he succeeded, where his colleagues had failed, in closing a fistula for the Duke of Chaulnes. As a consequence, Peyronie was called to Paris in 1707 at the instance of King Louis XIV, where he was appointed as surgeon-in-chief at the Hôpital de la Charité. At this time he also taught anatomy at the amphitheatre de Saint-Come and at the Jardin-du-Roi. Peyronie was especially skilful at operating on hernia and on the gut and he was eventually recognised as the foremost surgeon in France. In 1717 he became surgeon to King Louis XV and in 1721 he was elevated to the nobility.

In addition to his vast professional talents, Peyronie had considerable personal charm and an engaging bedside manner. Despite his accumulated wealth, he preferred a modest lifestyle and he was known for his sympathy with the impoverished. Peyronie was devoted to the interests of his fellow surgeons, and in 1731, in collaboration with his friend and former mentor, Mareschal, he used his influence with the King in order to establish a Royal Academy of Surgery. In this way he was able to define and to separate the profession of surgery from that of the barber.

Peyronie died in Versailles on 25 April 1747 at the age of 70 years. He bequeathed his library and an estate to the association of Parisian surgeons, while the surgeons of Montpellier received two mansions in the main street and a large sum of money for the construction of a lecture theatre.

# Nomenclature

I N 1743 Peyronie published a review of ejaculatory dysfunction, mentioning the fibrosing disorder that now bears his name. Several articles on the pathogenesis of penile fibrosis followed, and names which were applied to the condition included "cavenitis fibrosa", "fibrous sclerosis of the penis", "induratio penis plastica" and "primary indurative cavernositis". Van Buren and Keyes gave an account of the disorder in 1881, in a treatise on surgical disorders of the genitourinary system, and the former's eponym enjoyed transient recognition, although it is now rarely used.

The aetiology of Peyronie disease remains uncertain, although there is an association with Dupuytren contracture (*see The Man Behind the Syndrome* p. 47). Familial aggregation has also been recognised. In 1982 Wilma Bias and her colleagues at the Johns Hopkins Hospital, Baltimore, gave details of pedigree analyses of 3 affected kindreds and took the standpoint that Peyronie disease was a male-limited autosomal dominant disorder. In the 6th edition of *Mendelian Inheritance in Man* McKusick (1983) commented that an anonymous colleague had suggested that Peyronie disease was sex-linked, with reduced penetrance. Chromosomal defects in Peyronie disease were documented in 1991 by Guerneri and colleagues; heterogeneity seems possible but the precise pathogenetic mechanism is yet to be elucidated.

# References


Bayle ALJ, Thillaye AJ (1855) Biographie Médicale par ordre chronologique. 2: 194–195 A Delahaye, Paris

Bias WB, Nyberg LM, Hochberg MC, Walsh PC (1982) Peyronie's disease: a newly recognized autosomal dominant trait. Am J Med Genet 12: 227–235

Guerneri S, Stioui S, Mantovani F, Austoni E, Simoni G (1991) Multiple clonal chromosome abnormalities in Peyronie's disease. Cancer Genet Cytogenet 52: 181–185

Hirsch A (1931) *Biographisches Lexicon*. 2nd edn, Vol 3: 677–678

Peyronie F de la (1743) Sur quelques obstacles qui s'opposent à l'éjaculation naturelle de la semence. Mém de l'Acad Roy de Chir 1: 318–333

Van Buren WH, Keyes EL (1881) *A practical treatise on the surgical disease of the genito-urinary organs, including syphilis*. 2nd edn. Appleton, New York, pp. 81–83

# REIFENSTEIN, Edward Conrad

*(1908–1975)*

*Courtesy:* Professor John Opitz, Montana, USA

R EIFENSTEIN syndrome, or partial androgen insensitivity, is an X-linked form of male pseudohermaphroditism, which is characterised by hypogonadism, hypospadias, gynaecomastia, infertility and a normal XY karyotype.

# Biography

R EIFENSTEIN was a North American investigative endocrinologist. He made many contributions to the development of his speciality.

Edward Conrad Reifenstein was born in 1908 into a medical family in Syracuse, New York. His father was a practitioner in internal medicine and psychiatry, who remained active until his 90th year, and his 2 brothers were also medically qualified.

Reifenstein obtained his medical degree with distinction at the University of Syracuse in 1934. After internship and residency training, he specialised in psychiatry and undertook research into the use of amphetamine in the management of alcoholism. Reifenstein wished to pursue a career in research but felt under an obligation to join his father in his practice. This decision was strengthened by financial constraints imposed by his marriage, while still a medical student, and he joined his father's practice in 1937. Reifenstein retained his intellectual curiosity, and in 1939 he attended a 2-week course in endocrinology given in Boston by Fuller Albright. At that time endocrinology was a relatively new discipline and, excited by the future prospects, Reifenstein applied for postgraduate training with Albright. He was duly accepted and in the words of his biographers, Forbes and Bartter "That is how Ed's keen mind and his great capacity for hard and well-organised work came to be applied in a most fortunate milieu where they soon resulted in excellent and sustained achievement".

Reifenstein quickly gained both experience and status in Albright's department and after the outbreak of World War II he was involved in the investigation of the metabolic aspects of convalescence, including bone and wound healing. Reifenstein published extensively with Albright and his colleagues in these and related fields. He also took on the responsibility of editing the proceedings of the Macy Foundation meetings, at which endocrinologists from Baltimore, New York and Boston discussed their research findings. After the end of World War II, Reifenstein became chief of the endocrine unit at the Sloan-Kettering Institute and in 1950 accepted an appointment as director of a division of medical research in the pharmaceutical industry. His background in endocrinology, especially adrenal and gonadal function, were central to his success in this area.

In his personal life Reifenstein enjoyed mountaineering, the piano and his home. He retired in 1974 and died the following year. In his obituary, Forbes and Bartter stated "He remained as true to his principles as to his friends, and he commented that he was almost glad to be retiring from a profession which had grown too large for close teamwork and was tending to lose the disinterested devotion that he had found in, and given to, academic medicine. He also welcomes retirement, because he hoped it would give him more time to work. He envisioned remaining at his desk, reading and writing and serving as an editor and consultant. He was accustomed to being a pillar of strength, reliability and integrity in these capacities; by his many friends and colleagues he will be remembered and he will be missed."

# Nomenclature

S HORTLY after World War II, while working in Fuller Albright's laboratory in Boston, Reifenstein encountered a patient with an unusual gonadal abnormality. Reifenstein followed up the family and in 1947 he documented the disorder, using the term "hereditary familial hypogonadism". The same condition was reported in 1957, in the French literature by Gilbert-Dreyfus and colleagues. Bowen and colleagues in 1965 used Reifenstein's eponym in conjunction with the phrase "hereditary male pseudohermaphroditism" when they documented the phenotypic features of affected males in a large family. In 1974 Wilson and colleagues restudied the same family and designated the condition "familial incomplete male pseudohermaphroditism type I, Reifenstein syndrome", in order to distinguish this X-linked disorder from "type II", which is an autosomal recessive trait. The title "partial androgen insensitivity" was introduced in 1979 by Aiman and colleagues, in conjunction with the eponym. Reifenstein's name has been employed in further articles, and its use is firmly established.

# References


Aiman J, Griffin JE, Gazak JM, Wilson JD, MacDonald PC (1979) Androgen insensitivity as a cause of infertility in otherwise normal men. New Engl J Med 300: 223–227

Bowen P, Lee CNS, Migeon CJ, Kaplan NM, Whalley PJ, McKusick VA, Reifenstein EC (1965) Hereditary male pseudohermaphroditism with hypogonadism, hypospadias and gynecomastia: Reifenstein's syndrome. Ann Intern Med 62: 252–270

Forbes AP, Bartter FC (1976) Edward C Reifenstein, Jr. Recent Progress in Hormone Research 32: xiii–xvii

Gilbert-Dreyfus S, Sebaoun CA, Belaisch J (1957) Étude d'un cas familial d'androgynoidisme avec hypospadias grave, gynecomastie et hyperoestrogenie. Ann Endocr 18: 93–101

Reifenstein EC (1947) Hereditary familial hypogonadism. Recent Progr Horm Res 3: 224–225

Wilson JD, Harrod MJ, Goldstein JL, Hemsell DL, MacDonald PC (1974) Familial incomplete male pseudohermaphroditism type 1: evidence for androgen resistance and variable clinical manifestations in a family with the Reifenstein syndrome. New Engl J Med 290: 1097–1103

# RETT, Andreas
*(b. 1924)*

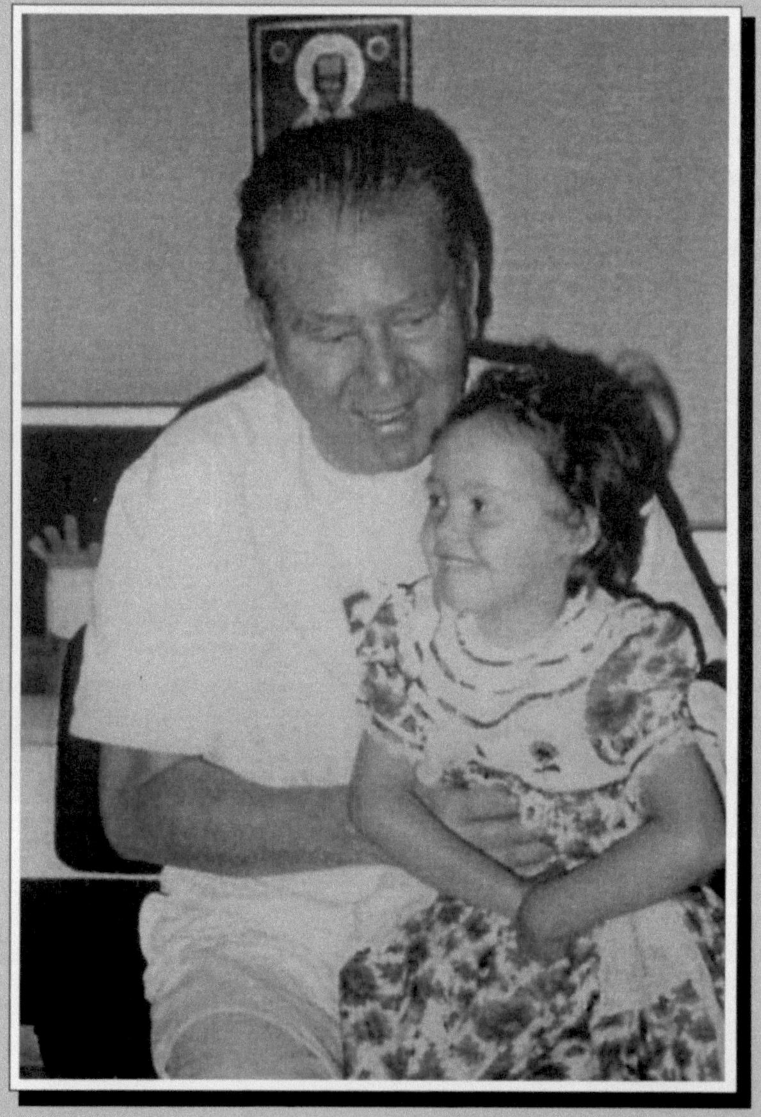

R ETT syndrome is a severe progressive disorder which comprises mental retardation and characteristic "hand-washing" mannerisms, with onset in early childhood. Only females are affected, and it is possible that the disorder is an X-linked dominant trait with male lethality.

# Biography

R ETT is a contemporary Austrian physician, with a special interest in the care of the mentally handicapped.

Andreas Rett was born on 2 January 1924, in Fürth, Bavaria, and received his schooling in Innsbruck. He achieved scholastic distinction and entered the medical school of the University of Innsbruck but his studies were interrupted by conscription into the German Navy. After twice being wounded, Rett was discharged from military service and in 1945 recommenced his medical studies at Innsbruck, qualifying in 1949. Rett then trained in Paediatrics at the Preyersches Kinderspital, Vienna, and under Guido Fanconi in Zurich. In 1955, after completion of his specialist training, Rett received an appointment as head of the facility for mentally retarded children at a home for the aged in Lainz, Austria where he delineated the disorder which bears his name. In 1967 he was appointed as lecturer in neurology and paediatrics at the University of Vienna, and as head of the Ludwig Boltzmann Institute for Research in Brain Disordered Children. Rett was promoted to the rank of associate Professor at the University of Vienna in 1973.

Rett was devoted to his patients and their welfare was preeminent in his scale of priorities. His personal qualities were exemplified when the President of the International Rett Syndrome Association, Kathy Hunter, described him as a "charming, charismatic, loving and sensitive man". In addition to his efforts for his patients, Rett found the time to publish more than 250 medical articles, mostly in the German literature. His contributions have been recognised by several awards, including the Gold Medal of Honor of the City of Vienna, which he received in 1984 and the Distinguished Achievement Award of the International Association for the Scientific Study of Mental Retardation in 1990.

Rett underwent open heart surgery in 1986 and thereafter suffered a stroke. Despite these problems, he was able to continue with his medical activities and in 1995, although his health was failing, he was still taking an interest in the condition which so deservedly bears his name.

# Nomenclature

A LTHOUGH the Rett syndrome is comparatively common, recognition was delayed until long after many much rarer disorders had been delineated. In an unpublished letter to the Rett syndrome association, written on 7 June 1985, and quoted by Rett's biographer Richard H. Haas, he described the events which led to his monumental discovery. "One day in the spring of 1965, two mothers holding their children on their laps were sitting in the waiting room. Both children were swaying and the mothers held their arms. Both children, who were treated for epileptic seizures, were well known to me. That morning I passed by them repeatedly and, incidentally, the mothers let go of their children's arms. At once, the children put their hands together and started nearly identically-looking washing movements. I asked the mothers not to stop these movements again, and was startled at the similarities. It was the same gaze, same facial expression, same weak muscles and the same stereotyped movements of their hands."

In 1966, Rett published his observations in the German literature, but unfortunately his article received little attention. He then made a film showing the characteristic appearance and mannerisms of affected persons. The disorder received recognition in Europe and in 1972 it was described in detail under Rett's name by Leiber and Olbrich in their catalogue of syndromes. Nevertheless, wider acceptance was delayed until 1983, 17 years after Rett's original description, when Hagberg and colleagues reported 35 affected girls in an article in the English literature. Rett's name was used in the title of this article and full acknowledgement of his seminal observations was made. Thereafter the syndrome attracted universal attention and Rett's name was perpetuated eponymously.

By 1986, it was estimated that there were at least 8,000 girls with the Rett syndrome in the USA alone, and at this time, a special workshop on the disorder was sponsored by the Kennedy Institute for the Handicapped Children, Baltimore, USA. Rett was a member of the organising committee, and in his opening historical and general review, he described the background to his observations. Rett closed his address with a poem *Portrait of a Rett's Child* by Gail Smith:

> *bright eyes*
> *sweet smile*
> *clasped hands*
> *unique style*
>
> *special child*
> *secret mind*
> *loving heart*
> *one of a kind*
>
> *so we thought*
> *but 'tis not true*
> *now we find she is one of a few"*

The Proceedings of the Rett Syndrome Symposium were published in 1986 in a supplement to the American Journal of Medical Genetics. International diagnostic criteria were established in 1988 and Rett Syndrome Associations were set up in several countries. Interest was sustained, and in 1994 a meeting entitled *Rett syndrome, from gene to gesture* was held at the Royal Society of Medicine, London. It is now estimated that at the age of 14 years, one in 10,000 females is affected. Despite the high frequency of the disorder, and the intensity of the investigations which have been undertaken, the pathogenesis of the Rett syndrome remains enigmatic.

National Rett Syndrome Associations have been established in over 30 countries. These associations, run mostly by parents, not only aim to contribute to the dissemination of knowledge about Rett syndrome but offer valuable support for parents, families and all persons concerned with this disorder.

# References


Haas RH (1990) Andreas Rett. In: Ashwal S (ed) *The Founders of Child Neurology*. Norman Publishing, San Francisco, pp. 822–826

Hagberg B, Aicardi J Dias K, Ramos O (1983) A progressive syndrome of autism, dementia, ataxia, and loss of purposeful hand use in girls: Rett syndrome: report of 35 cases. Ann Neurol 14: 471–479

Kerr A, Corbett J (1994) Rett syndrome: from gene to gesture. J Roy Soc Med 87: 562–565

Leiber B, Olbrich G (1972) *Die klinischen Syndrome: Rett' Syndrom*. Urban & Schwarzenberg, München, pp. 763–764

Rett A (1966) *Über ein cerebral-atrophisches Syndrom bei Hyperammonämie*. Brüder-Hollinek, Wien

Rett A (1986) Rett syndrome: history and general overview. Am J Med Genet 24, Suppl 1: 21–25

Rett A (1990) *Kinder in unserer Hand. Ein Leben mit Behinderten*. Orac Buch- und Zeitschriftenverlag GesmbH. Wien

# RILEY, Conrad M.

*(b. 1913)*

R ILEY-DAY syndrome, or familial dysautonomia, manifests with defective tearing, skin blotching, excessive sweating, postural hypotension and emotional lability. Inheritance is autosomal recessive and the condition predominates in persons of Ashkenazi Jewish stock.

## Biography

R ILEY is an American paediatrician with a special interest in renal disease.

Conrad M. Riley was born in Worcester, Massachusetts, USA in 1913 and graduated from Yale College in 1934. He qualified in medicine at Harvard in 1938 and his postgraduate medical training in internal medicine and paediatrics took place at the Presbyterian and Babies Hospital in New York City and Children's Hospital in Boston. Riley served in the Medical Corps of the US Naval Reserve from 1942 to 1946, and after the armistice, he returned to the Babies Hospital in New York City and joined the department of paediatrics at Columbia University.

While at the Babies Hospital, Riley pursued his special interest in nephrology in addition to his general clinical responsibilities. It was during this period that he encountered the condition which now bears his name. In collaboration with a psychiatric social worker and a paediatric psychiatrist, Riley established a group for the parents of affected children. The group provided valuable support for the parents and gave their medical practitioners a greater understanding of the syndrome.

During his time in New York, Riley maintained his interest in renal disease and published several papers on childhood nephrosis. In 1960 he moved to the department of paediatrics of the Colorado University School of Medicine, Denver, and from 1961 until 1966 he was chairman of the department of preventive medicine. Riley remained active in both departments until 1972 when he became associate dean for admissions. In 1977 he became emeritus professor but continued his relationship with the medical school, serving on the ethics committee and on a committee to review research involving human beings.

Riley's principle avocation has been the Denver Zoo, and he has served on the Board of Trustees of the Denver Zoological Foundation since 1973 and as its president from 1980 to 1985. Over the years it has grown to become a world-class institution.

## Nomenclature

I N 1949, at the Babies Hospital, New York, Riley encountered several children presenting with a condition that in those days was known as "cyclic vomiting". Among these were the early cases that were subsequently grouped together as a single clinical entity which came to be termed "familial dysautonomia". Dr Riley's colleagues, Richard Day, David Greeley and William Langford, each contributed details about separate patients and together they agreed that the initial 5 cases represented a specific genetic entity. In their article which was entitled *Central autonomic dysfunction with defective lacrimation*, they stated "In the course of the past 10 years four children have appeared at Babies Hospital with symptoms so puzzling as to defy exact diagnosis yet so similar as to constitute a clinical entity. A fifth patient with the same condition has been studied at The New York Hospital. The common features are:

1. undue reaction to mild anxiety characterized by excessive sweating and salivation, red blotching of the skin and transient but marked arterial hypertension, and 2. constantly diminished production of tears. Search among textbooks and current literature has failed to reveal a description of a syndrome corresponding to the one described". Since the publication of this article, the literature on the Riley-Day syndrome has become extensive. Recent interest has focussed on the elucidation of the basic defect and in 1993 Blumenfeld et al. assigned the gene for familial dysautonomia to chromosome 9. (For further details of the nomenclature *see* Day, p. 53.)

Riley's name has been associated with that of Schwachman (*see The Man Behind the Syndrome* p. 227) as a designation for a condition in which osseous changes were combined with hyperreflexia. Apart from their original description of two affected children, there have been few, if any additional reports, and the syndromic status of this disorder is not yet fully established.

## References


Blumenfeld A, Slaugenhaupt SA, Axelrod FB, Lucente DE, Maayan C, Liebert CB, Ozelius LJ, Trofatter JA, Haines JL, Breakefield XO, Gusella JF (1993) Localization of the gene for familial dysautonomia on chromosome 9 and definition of DNA markers for genetic diagnosis. Nature Genet 4: 160–164

Riley CM (1952) Familial autonomic dysfunction. JAMA 149: 1532–1535

Riley CM, Schwachman H (1943) Unusual osseous disease with neurologic changes; report of two cases. Am J Dis Child 66: 150–154

Riley, CM, Day RL, Greeley DM, Langford WS (1949) Central autonomic dysfunction with defective lacrimation. I. Report of five cases. Pediatrics 3: 468–478

# ROBINOW, Meinhard
*(b. 1909)*

R OBINOW, or foetal face syndrome, comprises an unusual facies, stunted stature, short forearms and genital hypoplasia. Autosomal dominant and autosomal recessive forms have now been well documented, the 2 forms being phenotypically distinguishable.

# Biography

R OBINOW is a well-known paediatrician in the USA. Over a 40-year period he made numerous contributions to the field of syndrome delineation.

Meinhard Robinow was born on 19 May 1909 in Hamburg, Germany, where his father was a lawyer. He attended the Humanistisches Gymnasium and then studied medicine at the Universities of Munich, Heidelberg, Berlin and Hamburg graduating in 1934. Shortly after qualification he moved to the USA to avoid Nazi persecution and, after a rotating internship and a year of cardiology in Chicago, he trained in paediatrics in Augusta, Georgia. Robinow then spent 3 years investigating the growth of normal children at the Fels Research Institute, Yellow Springs, Ohio.

During World War II Robinow served in the US Army Medical Corps in Britain, France, Okinawa and Japan. After demobilisation he commenced private paediatric practice in Yellow Springs, Ohio, and continued in this activity until 1975. In this phase of his career, Robinow described the syndrome which bears his name. He also became increasingly involved with birth defects and returned to academic life at the University of Virginia where his major research interests were congenital anomalies of the skeleton and syndrome delineation. He subsequently moved to Wright State University and the Children's Medical Center in Dayton, Ohio, recommencing part-time paediatric practice when he reached compulsory retirement age.

In an encomium published in the *American Journal of Medical Genetics* in 1995, it was stated "At the age of 86, Meinhard Robinow, still active as a clinical geneticist and passionate horticulturist, is entitled to look back on an exceptionally interesting and fruitful life with great satisfaction. What a blessing for the fields of growth analysis, syndromology, and clinical genetics in general that Meinhard did not fall a victim of Nazi persecution, but, like so many others was accepted warmly in the New World, where he was destined to develop his great skills as a pediatric geneticist and diagnostician." The encomium concluded with the comment "It is a reflection of a basic humility to refer to oneself as an 'amateur' after having made so many recognized, substantial contributions to the field of clinical genetics. What is most admirable about the life of Meinhard Robinow is that this professional humility is coupled with an evident love of life, an infectious enthusiasm, and a devotion to hard work, which allowed him to make contributions of such importance that the name Robinow shall not perish from the annals of clinical genetics."

In his leisure time Robinow enjoys tennis and squash. He is also a keen horticulturist and has an encyclopaedic knowledge of the flora and fauna of Ohio. Robinow married Beatrice Wilt, an Army nurse, in 1944 and the couple have 3 children. In 1995 they were living on their farm in Ohio, producing flowers, fruit and vegetables. Robinow also continues a small private practice.

# Nomenclature

I N 1963, while in private practice in Yellow Springs, Ohio, Robinow was asked to see a newborn infant for "ambiguous genitalia". In describing the history and the condition which now bears his name, Robinow stated: "The infant had short limbs, an unusual facies and a minute phallus with a central urethra, completely buried in the suprapubic fat. A buccal smear indicated that male sex (later confirmed by chromosome studies and a testicular biopsy). When I saw the mother it became obvious to me that she had the same condition. I phoned F.N. Silverman, the well known paediatric radiologist who was then located in nearby Cincinnati, Ohio and told him that I thought I had discovered a new syndrome. He was at first skeptical but agreed after he saw the patient and his mother. Two years later the mother gave birth to a similarly affected girl and after two and a half more years, to another affected boy."

The members of this family had moderate dwarfing, with mesomelic limb shortening, hypertelorism, a bulging forehead and a depressed nasal bridge. The males had micropenis and there was hypoplasia of the clitoris and labia minora in the females. The maternal grandmother was similarly affected and family photographs strongly suggested that members of 3 preceding generations had also been affected.

Robinow continued "Strange as it may seem now, the characteristic radiographic features of achondroplasia had not yet been well delineated. Various skeletal dysplasias were called 'atypical achondroplasias'. It therefore seemed necessary to stress that our syndrome was neither achondroplasia nor any other known chondrodysplasia.

Robinow, Silverman and Smith published an account of this family in 1969 and tabulated the differences between the disorder and achondroplasia. In 1973 Robinow reviewed the condition in the *Birth Defects Compendium*, and since the face resembles that of a human foetus at 8 weeks gestation, introduced the title "fetal face syndrome". Wadlington et al. (1973) used the eponym "Robinow" in the titles of their papers and thereafter the eponym became established.

Autosomal recessive forms of the Robinow syndrome have been described by several authors (Wadlington et al. 1973; Balci et al. 1991) and in 1993 Robinow distinguished the two forms phenotypically. Patients with the AR form tend to have more severe forearm brachymelia and vertebral anomalies, more digital defects and a more triangular mouth, while the AD form has mildly stunted-to-normal stature, almost no vertebral segmentation defects, mild forearm brachymelia and minor digital anomalies.

In addition to the foetal face syndrome Robinow's name is also used eponymously in conjunction with that of Sorauf for an autosomal dominant craniosynostosis-bifid hallux syndrome, which is also known as acrocephalopolysyndactyly. (Robinow and Sorauf 1975; Carter et al. 1982).

# References


Balci S, Erçal B, Say B, Turner M (1991) Robinow syndrome: 14 cases with special emphasis on dermatoglyphics and hand malformations (split hand at three of them). Int Congr Hum Genet, p. 127

Carter CO, Till K, Fraser V, Coffey R (1982) A family study of craniosynostosis, with probable recognition of a distinct syndrome. J Med Genet 19: 280–285

Garn SM, Kelly TE, Opitz JM (1995) Meinhard Robinow: An Appreciation. J Med Genet 59: 4–7

Robinow M (1973) Fetal face syndrome. In: Bergsma D (ed) *Birth Defects Atlas and Compendium*. Williams & Wilkins, Baltimore, pp. 410–411

Robinow M (1993) The Robinow (fetal face) syndrome: a continuing puzzle. Clin Dysmorph 2: 189–198

Robinow M (1995) Living history – autobiography: pediatric genetics in a New World. J Med Genet 59: 8–13

Robinow M, Sorauf TJ (1975) Acrocephalopolysyndactyly, type Noack, in a large kindred. Birth Defects Orig art Ser XI(5): 99–106

Robinow M, Silverman FN, Smith HD (1969) A newly recognized dwarfing syndrome. Am J Dis Child 117: 645–651

Wadlington WB, Tucker VL, Schimke RN (1973) Mesomelic dwarfism with hemivertebrae and small genitalia (the Robinow syndrome). Am J Dis Child 126: 202–205

# ROMBERG, Moritz Heinrich

## *(1795–1873)*

*From*: Viets H.R. (1953) In: Webb Haymaker (ed) The Founders of
Neurology, 1st edn

*Courtesy*: Charles C Thomas, Springfield, Illinois
Army Medical Library, Washington, DC

R OMBERG, or Parry-Romberg syndrome, is characterised by slowly progressive atrophy of the soft tissues of one side of the face, associated with trigeminal neuralgia and involvement of the eyes and hair.

# Biography

R OMBERG was a leading German neurologist of the nineteenth century and the author of the first modern classification of neurological disorders.

Moritz Heinrich Romberg was born in 1795 in Meiningen and qualified in medicine at the University of Berlin in 1817. His doctoral thesis, which was written on the subject of "congenital rickets", contained a comprehensive description of achondroplasia. In 1820 Romberg undertook a brief period of postgraduate training in Vienna, under Johann Peter Frank, with whom he developed a lasting friendship. Frank was a pioneer in public health and an expert on disorders of the spinal cord. He had a profound influence on Romberg and was responsible for his career being directed towards neurology. After his return from Vienna, Romberg became medical officer to the paupers of Berlin and during the epidemics of 1830 and 1836, he was in charge of a cholera hospital. In 1845 he was elevated to the Chair of Medicine and directorship of the Royal Policlinic Institute at the Friedrich Wilhelm University, Berlin.

Romberg's academic reputation was established when he published his 3-volume *Lehrbuch der Nervenkrankheiten* (1840, 1843, 1843) in which he discussed the physiology of neurological function, endeavoured to define clinico-pathological correlations and constructed a nosology of nervous system disorders. He had a firm belief in his own abilities and in the preface to this book he castigated his predecessors for the inadequacy of their contributions: "The blame lies in a measure with the distinguished members of our profession who have been deterred by a fear that pathological investigations would fail to cope with the advanced state of physiological inquiry; in others, the fault is to be attributed to that mental indolence, which gives the preference to the easy path of tradition, and with foolish scepticism rejects everything that is new."

Romberg's work had considerable influence on neurological concepts, especially after publication of a version which had been translated into English, under the sponsorship of the Sydenham Society. His enthusiasm for this field had been generated by the researches of Sir Charles Bell (*see* p. 21) whom he acknowledged in his preface. Romberg's nosology is now outmoded but in its time it represented the first real attempt to systematise neurological disease.

The tradition of medical eponymy in Romberg's family was maintained when his nephew, Eduard Henoch, described the form of purpura which now bears his name. Romberg died of heart disease in Berlin on 16 June 1873 at the age of 78 years.

# Nomenclature

T HE first mention of the condition which bears Romberg's eponym was made by Caleb Hillier Parry (1755–1822). Parry was a general practitioner in Bath, England, and his medical writings were published posthumously in 1825.

In 1846 Romberg gave a complete description of the disorder in his *Klinische Ergebnisse*, and in 1871 Eulenberg introduced the modern descriptive title "progressive hemifacial atrophy". The dramatic manifestations of the condition have prompted numerous case reports, and in addition to the descriptive designation, the names of Romberg and Parry have been used together and separately. The disorder is not uncommon, and in 1963 Rogers was able to review 772 cases. An association with facial linear scleroderma (scleroderm en coupe de sabre) has been postulated and antinuclear antibodies have been incriminated. Although there have been some instances of apparent generation-to-generation transmission, in keeping with autosomal dominant inheritance, the majority of affected persons have been sporadic. In the present state of knowledge, the aetiology and pathogenesis of the condition remain uncertain.

In addition to the progressive hemifacial atrophy, Romberg's name is used in conjunction with that of John Howship (1781–1841) in the context of neuralgic leg pain due to obturator hernia. His eponym is also occasionally conjoined with that of Karl Westphal (1833–1890) for a form of familial periodic paralysis. At the present time he is best known for the "Romberg sign" which is demonstrated when patients with locomotor ataxia are unable to maintain a stationary stance when standing with their eyes closed. (*See* Parry, p. 129 for further details of the nomenclature of the Parry-Romberg syndrome.)

# References


Eulenberg A (1871) *Lehrbuch der Functionellen Nervenkrankheiten auf Physiologischer Basis*. A. Hirschwald, Berlin, p. 712

Parry CH (1825) *Collections from the unpublished medical writings of the late Caleb H Parry*. Underwoods, London.

Rogers BO (1963) Progressive facial hemiatrophy: Romberg's disease: a review of 772 cases. Proceedings of the 3rd International Conference on Plastic Surgery. Excerpta Med Int Cong Series 66: 681–686

Romberg MH (1817) *Inaugural Dissertation on Congenital Rickets*. Karl August Platen, Berlin

Romberg MH (1846) *Trophoneurosen, Klinische Ergebnisse*. Förstner, Berlin, pp. 75–81

Romberg MH (1853) *A Manual of the Nervous Diseases of Man*. Translated by H Sieveking. Sydenham Society, London

# RUSSELL, Alexander
*(b. 1914)*

USSELL dwarfism is characterised by low birth weight, delayed skeletal development and slight stature. The face is triangular with a down-turned mouth and a high bossed forehead.

# Biography

USSELL is an academic paediatrician who had a distinguished career in Britain and Israel.

Alexander Russell was born in Britain and after demonstrating early scholastic ability, qualified in medicine at the University of Durham at the age of 21 years. He undertook junior hospital appointments in Newcastle-upon-Tyne, working with Professor Beattie, Gray Turner and Spence. The latter was also the head of the British Medical Research Council at the time, and Russell was inspired by his humanisation of child care.

Russell's career was interrupted by World War II and he spent 6 years as a medical officer in the Royal Air Force. During his war service he was mentioned in despatches for his recognition of carbon monoxide poisoning in the poorly ventilated rear cockpits of Whitley bombers. He was also awarded the Order of the British Empire for his innovative use of DDT in the Middle East in programmes of prevention of Kala Azar by destruction of the sandfly disease vector.

Russell resumed his academic career after the armistice and in 1948 he was awarded his doctorate, with distinction, for a thesis on the "Diencephalic Syndrome of Hyperkinetic Emaciation". He continued with his life-long orientation towards the community and preventative aspects of child care, and in 1950 he became assistant to Professor Alan Moncrieff at the Great Ormond Street and Queen Elizabeth Children's Hospitals, London. At the latter he founded the first paediatric endocrine unit in Britain, where he was able to focus upon growth, metabolic and neuro-endocrinological disorders. Together with his research assistant, S. Butler, he also established the first clinical cytogenetic unit in a British children's hospital.

In 1966 Russell was called to the newly established Chair of paediatrics and Child Care at the Hadassah Medical Centre at the Hebrew University of Jerusalem. In this role he was instrumental in the development of a network of mother and child clinics, which combined preventitive, diagnostic, curative and rehabilitation services. The early detection of handicap and integration of affected children with their normal peers was central to this initiative. In 1984 Russell's career was crowned by his election to the Presidency of the International College of Paediatrics.

# Nomenclature

N 1954 Russell presented 5 children with prenatal growth retardation at the Royal Society of Medicine, London. His article, which appeared in the society's transactions, was entitled *Intrauterine dwarfism, with craniofacial dysostosis, disproportionately short arms and other anomalies*. In the preceding year, Silver of Denver and his colleagues (1953) had reported 2 children with a syndrome of "congenital hemihypertrophy, shortness of stature and elevated urinary gonadotropins". Controversy developed as to whether any of Russell's patients had the condition previously described by Silver, and whether or not the Russell syndrome and the Silver syndrome were separate entities. In 1964 Silver published a second article, in which he emphasised asymmetry as a syndromic component. The question of syndromic identity remained unsettled until 1969, when Tanner and Ham suggested that the designation "Russell syndrome" should be used for the form of the condition in which asymmetry was lacking, while the term "Silver syndrome" should be used when the patient had asymmetry. This viewpoint does not reflect the historical situation, as Russell certainly drew attention to asymmetry in his original article.

Against this background, the continued use of the conjoined eponym is appropriate. The aetiology of the Russell-Silver syndrome is still uncertain and it has not been possible to reproduce the original observations of Silver and colleagues concerning abnormal levels of urinary gonadotrophins. A few familial instances have been documented and in a single family, Partington (1986) recognised X-linked transmission (*see The Man Behind the Syndrome* Silver, p. 228).

# References


Partington MW (1986) X-linked short stature with skin pigmentation: evidence for heterogeneity of the Russell-Silver syndrome. Clin Genet 29: 151–156

Russell A (1954) A syndrome of "intra-uterine" dwarfism recognizable at birth with cranio-facial dysostosis, disproportinately short arms and other anomalies (5 examples). Proc Roy Soc Med London 47: 1040–1044

Silver HK (1964) Asymmetry, short stature and variations in sexual development: a syndrome of congenital malformations. Am J Dis Child 107: 495–515

Silver HK, Kiyasu W, George J et al. (1953) Syndrome of congenital hemihypertrophy, shortness of stature and elevated urinary gonadotropins. Pediatrics 12: 368–376

Tanner JM, Ham TJ (1969) Low birthweight dwarfism with asymmetry (Silver's syndrome): treatment with human growth hormone. Arch Dis Child 44: 231–243

# SAETHRE, Haakon

*(1891–1945)*

*Courtesy*: Professor S. Refsum, Norway

S AETHRE-CHOTZEN syndrome, or acrocephalo-syndactyly type III, is a mild autosomal dominant disorder in which stenosis of the coronal sutures leads to a characteristic cranio-facial configuration. Partial cutaneous syndactyly and minor digital abnormalities may also be present.

# Biography

S AETHRE was a prominent Norwegian neuro-psychiatrist. His successful career was abruptly terminated in 1945 when he was murdered in Nazi reprisals.

Haakon Saethre was born in Bergen in 1891 and graduated in medicine from the University of Oslo in 1918. He trained in neurology and psychiatry at the State Hospital and in 1933 he was appointed as chief physician to the psychiatric department of the City of Oslo Hospital, Ullevål. Saethre also served as medical consultant to the Oslo City Child Committee, where he made proposals for the establishment of special child psychiatry clinics, and he also maintained a large private practice.

Saethre was an outstanding clinician and administrator and he quickly made an impact in his field. He represented Norway at several international meetings and in 1938 he was president of the Scandinavian Congress of Psychiatry. Saethre was also honorary president of the Norwegian Society of Mental Hygiene for several years. He combined his clinical work with initiatives for the care of the sick in the community and he was regarded by his colleagues as a knowledgeable and devoted doctor and a skilled teacher.

In the early stages of his career Saethre investigated clinical problems, including craniostenosis, multiple sclerosis and tabes dorsalis. In particular, his studies of the cerebrospinal fluid in neurosyphilis which revealed correlations with defined regimes of treatment, attracted international interest. Later he focused on chronic alcoholism and the psychiatric effects of head injury.

Saethre had a strong sense of patriotism and he was implacably opposed to the Nazis, who occupied his country during World War II. He was active in the resistance movement, aiding Jews to escape to Sweden and sometimes concealing them by admission as patients to the hospital wards. In February 1945, 3 months before liberation, a German-appointed senior police officer was assassinated in Oslo by the Norwegian resistance force. As a reprisal, the Germans rounded up several prominent Norwegian civilians, including Saethre, who was arrested at his hospital. On the following morning he was shot and his body immediately cremated. His ashes were said to have been thrown into the Oslo Fjord.

Saethre's portrait was painted posthumously by his friend, Henrik Sorensen, a prominent artist. This portrait now hangs in the department of psychiatry, City of Oslo Hospital, Ullevål. Saethre was 54 years of age at the time of his death.

# Nomenclature

I N 1931 Saethre wrote review articles in the Scandanavian and German literature on the subject of oxycephaly. In these publications he documented a mother aged 32 years, and her 2 daughters, who had mild brachycephaly and asymmetry of the skull, with partial soft tissue syndactyly of the second and third fingers. In the following year Chotzen[1] reported a father and 2 sons with similar abnormalities. Several reports followed, resulting in some nosological confusion but the situation was clarified when numerical designations I to V were applied to the various forms of acrocephalosyndactyly.

In 1970 Bartsocas and colleagues documented an affected Lithuanian family in the USA, in which 10 persons in 3 generations had the condition which Saethre had described. The designation "acrocephalosyndactyly type 3, Chotzen's syndrome" was used in the title of this article. Saethre's name is now conventionally added, with priority in the conjoined eponym.

# References


Bartsocas CS, Weber AL, Crawford JD (1970) Acrocephalosyndactyly type 3: Chotzen's syndrome. J Pediat 77: 267–272

Chotzen F (1932) Eine eigenartige familiaere Entwicklungsstoerung (Akrocephalo-syndaktylie, Dysostosis craniofacialis und Hypertelorismus). Mschr Kinderheilk 55: 97–122

Obituary (1945) Br Med J. April: 536–537

Saethre M (1931) Ein Beitrag zum Turmschaedelproblem (Pathogenese, Erblichkeit und Symptomatologie). Dtsch Z Nervenheilk 119: 533–555

Saethre M (1931) Ueber den Turmschädel, seine Erblichkeit, Pathogenese und neuropsychiatrischen symptome. Acta Neuro Psychiat Scand 6: 405–427


---

[1]Dr F. Chotzen was a German psychiatrist who described the condition in 1932.

# SECKEL, Helmut P.G.

## *(1900–1960)*

*From*: J Pediatr (1960) October: 638

*Courtesy*: Mosby-Yearbook Inc, St Louis

S ECKEL syndrome is an autosomal recessive heterogeneous disorder characterised by proportionate dwarfism and a bird-like facies. Joint dislocations, cryptorchidism and mental retardation are variable features.

# Biography

S ECKEL was a German paediatrician who spent the major part of his career at the University of Chicago, USA during the middle years of the twentieth century.

Helmut Seckel was born on 16 May 1900 in Berlin. He came from a family of prominent academics, his father being professor of Roman law at the University of Berlin and a member of the Prussian Academy of Science. Seckel's younger brother achieved similar distinction as professor of the History of Art at the University of Heidelberg, where he had an international reputation for his knowledge of Japanese art.

After qualification in medicine in Berlin, Seckel undertook postgraduate training in paediatrics under Czerny, Moro and Kleinschmidt. He became a paediatrician at the University of Cologne but his career was interrupted when the Nazis came to power and in 1936 he fled to the USA. Seckel obtained an appointment in the Department of Pediatrics, University of Chicago Medical School, where he eventually succeeded to the Chair. He published extensively in the fields of infectious disease, endocrinology and child development, acquiring an international reputation for his contributions in these fields.

Seckel died in 1960 at the age of 59 years. In his obituary, his friend Grossman wrote "Dr Seckel will always be remembered as a scholarly man. In his academic pursuits he explored many fields: medical science, social science, art, philosophy, history and theology. His readings and discussions in the field of existentialism were wide and profound. Many felt he was old-fashioned because he respected the old German traditions, though he was appalled at the degradation to which the Nazi curse had led the German people. Yet his was an open mind, ever accepting of new ideas, the newest and most abstruse expressions of art, the newest advances in medical practice. He enjoyed the company of young people, colleagues and students, and was willing to learn from them. He was a quiet man, a gentle man, a dignified man, yet humble and kind."

# Nomenclature

I N October 1959 Seckel completed a monograph entitled *Bird-Headed Dwarfs: Studies in Developmental Anthropology, including Human Proportions.* This work, which was published in 1960, in the year of his death, contained a detailed account of 2 unrelated children whom he had studied in Chicago, together with a review of information concerning 13 additional nanocephalic dwarfs reported in the literature over a 200-year period. Amongst the affected persons whom Seckel discussed was Caroline Crachami, whose skeleton now reposes in the Hunterian Museum of the Royal College of Surgeons, London. Crachami, an exhibitionist known as the "Sicilian Fairy" was 49.5 cm in height when she died in 1824 at the age of 9 years.

The term "bird headed dwarf" had initially been introduced by Virchow, in the context of proportionate dwarfism with low birth weight, microcephaly, mental retardation, a pointed nose, micrognathia and other variable stigmata. Seckel's monograph generated renewed interest and articles which followed usually employed some form of the designation "Seckel's bird headed dwarfism".

In 1967 McKusick and colleagues documented the condition in 3 out of 11 siblings and suggested that inheritance was probably autosomal recessive. It is likely that the disorder is heterogeneous, although there is controversy concerning syndromic boundaries. Indeed, in an extensive account of historical details concerning Caroline Crachami, Bondeson (1992) questioned the nosological status of bird-headed dwarfism and of the Seckel syndrome.

In recent times the descriptive term "bird headed dwarf" has come to be regarded as being perjorative, and the simple designation "Seckel syndrome" is in general use. As an alternative, the name "microcephalic primordial dwarfism" is sometimes employed, with numerical designations to indicate sub-types.

# References


Bondeson J (1992) Caroline Crachami, the Sicilian Fairy: a case of bird-headed dwarfism. Am J Med Genet 44: 210–219

Majewski F, Goecke T (1982) Studies of microcephalic primordial dwarfism. I: Approach to a delineation of the Seckel syndrome. Am J Med Genet 12: 7–21

McKusick V A, Mahloudji M, Abbott MH, Lindenberg R, Kepas D (1967) Seckel's bird-headed dwarfism. New Engl J Med 277: 279–286

Seckel HPG (1960) *Bird-headed dwarfs: studies in developmental anthropology.* Charles C Thomas, Springfield, Illinois

Virchow R (1892) Vorstellung der Knaben Dobos Janos. Berl Klin Wschr 29: 517

# SPIELMEYER, Walther

## (1879–1935)

S PIELMEYER-VOGT disease, or neuronal ceroid-lipofucinosis type III, is an autosomal recessive neurodegenerative disorder of early childhood, characterised by progressive deterioration of vision and intellect.

# Biography

S PIELMEYER was a distinguished German neuro-histopathologist. He made many contributions to the understanding of the pathological basis of neurological disorders.

Walther Spielmeyer was born on 23 April 1879 in Dessau, where he was the youngest of several siblings. He initially contemplated a career in the Church but decided upon medicine and studied at Halle. While still a medical student he commenced his life-long interest in neuropathology, undertaking histological investigations of the brains of deceased psychotics.

After graduation Spielmeyer trained in psychiatry in Freiburg, under Hoche. He continued with his histopathological studies, being greatly influenced by Nissl of Heidelberg, and soon established a reputation as a highly promising young scientist. Spielmeyer was called to Munich to work under Kraepelin and in 1912 he succeeded Alzheimer as laboratory head. During World War I he worked in the field of peripheral nerve injuries, and in 1918 he was elevated to honorary professorial status. Spielmeyer was increasingly involved with establishing correlations between histological changes in the brain and neurological dysfunction, especially in relation to circulatory abnormalities. His research output was prodigious and in 1922 he published the first textbook on the histopathology of the nervous system. In 1928 Spielmeyer was appointed as director of histopathology at the new Kaiser Wilhelm Institute, which was financed by the Rockefeller Foundation. He remained in this post until his death in 1935.

Spielmeyer was a quiet, sophisticated man with a formal manner. He was an accomplished pianist and singer, and frequently entertained professional musicians in his home. Spielmeyer had a strong streak of obsessionalism and was a perfectionist in his academic activities. These traits were also manifest in minor eccentricities; for instance, he steadfastly refused to examine histological sections of neoplasms, and he would never use gold compounds in the staining of slides.

Towards the end of his life Spielmeyer's intellectual honesty and open criticism brought him into conflict with the Third Reich. In 1935 he contracted tuberculosis which was complicated by influenzal pneumonia and he died after a short illness.

# Nomenclature

I N 1908, while working in Freiburg, Spielmeyer investigated histological specimens from the brain of a child who had died from "amaurotic family idiocy". He recognised that the nervous tissue was infiltrated with a fatty substance and concluded that lipid metabolism was disturbed. At this time the amaurotic familial idiocies were an ill-defined group of disorders in which progressive visual and intellectual deterioration were the main features and Spielmeyer was the first to recognise the nature of the underlying abnormality. At the turn of the century the Batten brothers, Rayner and Frederick, had individually documented separate sets of affected siblings in England and in the British literature the name of Batten (see *The Man Behind the Syndrome* p. 13) was used for this condition. Amongst other early reports was an article written in 1904 by Marmaduke Stephen Mayou concerning familial cerebral degeneration and retinal changes, and another by Heinrich Vogt (1905) on family amaurotic idiocy (*see* p. 177). In 1908, Wolfgang Stock, a German physician, published an account of the same disorder and the double or triple eponym came into use. Later, these conditions were subdivided in terms of the age of onset, the variety studied by Spielmeyer being designated the "juvenile" form. A late infantile type now bears the eponyms of Jansky, a Czech physician, and Bielschowsky (*see* p. 25), while an adult form is known as Kufs[1] disease.

The nature of the metabolic defect has been elucidated in this group of disorders, and the biochemical designation "neuronal ceroid lipofuscinosis" is currently employed. Numerical designations have been given to the biochemical sub-types, with the retention of appropriate eponyms. Because of perjorative connotations, the original appellation "amaurotic family idiocy" has been discarded.

# References


Batten FE (1903) Cerebral degeneration with symmetrical changes in the maculae in two members of a family. Trans Ophthalmol Soc, UK 23: 386–390

Batten RD (1897) Two brothers with symmetrical disease of the macula, commencing at the age of fourteen. Trans Ophthalmol Soc, UK 17: 48–50

Berkovic SF, Carpenter S, Andermann F, Andermann E, Wolfe L (1988) Kufs disease: clinical features and forms. Am J Med Genet Suppl 5: 105–109

Kufs H (1931) Über einen Fall von spätester Form der amaurotischen Idiotie mit dem Beginn im 42 und Tod im 59 Lebensjahre in klinischer, histologischer und vererbungspathologischer Beziehung. Z Ges Neurol Psychiatrie 137: 432–448

Mayou MS (1904) Cerebral degeneration with symmetrical changes in the maculae, in 3 members of a family. Trans Ophthmol Soc UK 24: 142–145

Obituary (1935) Walther Spielmeyer. J Nerv Ment Dis 81: 724–725

Spielmeyer W (1908) *Klinische und anatomische Untersuchungen über einen besonderen Fall von amaurotischer Idiotie.* Nissls Beitr Nerv Geistes Krkh, Berlin

Stock W (1908) Über eine bis jetzt noch nicht beschriebene Form der familiär auftretenden Netzhautdegeneration bei gleichzeitiger Verblödung und über Pigmentdegeneration der Netzhaut. Klin Mbl Augenh 5: 225–244

Vogt H (1905) Über familiäre amaurotische Idiotie und verwandte Krankheitsbilder. Mschr Psychiat 18: 161–167


---

[1]Hugo Friedrich Kufs was a German neurologist whose name was applied to the late onset form of amaurotic familial idiocy (now late-onset neuronal ceroid-lipofuscinosis) after he documented 3 affected persons in 1925, 1929 and 1931. The nosology and natural history of Kufs disease was reviewed in detail in 1988 by Berkovic and colleagues.

# STARGARDT, Karl Bruno

*(1875–1927)*

S TARGARDT disease, or familial juvenile macular degeneration, is a heterogeneous disorder which presents in mid-childhood and slowly progresses to total blindness.

## Biography

S TARGARDT was an academic ophthalmologist in Germany. His career was curtailed by his early death a few years after his appointment to a chair at the University of Marburg.

Karl Stargardt was born in Berlin on 4 December 1875. He studied medicine in Heidelberg, Erlangen and Berlin, and qualified in Kiel in 1899. Stargardt trained in ophthalmology in Kiel under Völkers and was then appointed as acting head of the eye clinic at Strasburg. He subsequently became head of ophthalmology at Bonn, and in 1923, after the retirement of Bielschowsky (*see* p. 25), was called to the chair at the University of Marburg.

Stargardt made many contributions to the science of ophthalmology. In the early stages of his career he was involved in the study of the mechanisms and testing of light and colour perception. Later he worked in the field of ophthalmological bacteriology using the trypanosome to produce experimental keratitis. Stargardt also elucidated the histopathology of optic nerve damage due to neurosyphilis and was involved in the delineation of different forms of macular degeneration.

In 1923, shortly after his appointment to the chair of ophthalmology at Marburg, Stargardt developed nephritis and cardiac complications. Thereafter he was often confined to his bed and, although he was able to continue with his research to some extent, he died on 2 April 1927. Stargardt was held in high regard by his colleagues and it was generally agreed that his untimely death had blighted a fruitful academic career.

## Nomenclature

I N 1909 Stargardt wrote an account of 7 persons in 2 families in whom a progressive form of macular degeneration commenced in childhood and led to blindness. He observed that, although 2 male and female siblings were affected, the condition was confined to a single generation in each family. In this article he also depicted the retinal changes which are characteristic of the condition which bears his name. In 1917 Stargardt documented another affected person. The familial nature of the disorder was re-emphasised in 1935 by Wright and other reports followed.

In 1971 Deutman listed all publications on "Stargardt's disease" and by this time it was evident that several separate disorders were grouped together in this category. In the following year Krill and Deutman summarised their experience of almost 150 affected persons, reviewed the literature concerning the macular dystrophies and proposed syndromic identity for the specific condition which had been documented by Stargardt. They also suggested that the disorder previously named "fundus flavimaculatus" by Franceschetti in 1963 was the same entity.

The term "Stargardt disease" is now widely used, with "familial juvenile macular degeneration" and "juvenile hereditary disciform macular degeneration" as alternative descriptive designations.

Inheritance is usually autosomal recessive, although an uncommon autosomal dominant form has also been delineated. This latter condition was documented in 34 persons in 5 generations by Cibis and colleagues in 1980; the fundal appearances led to the suggested title "macular dystrophy with flecks".

## References

Cibis GW, Morey M, Harris DJ (1980) Dominantly inherited macular dystrophy with flecks (Stargardt). Arch Ophthalmol 98: 1785–1789

Deutman AF (1971) *The Hereditary Dystrophies of the Posterior Pole of the Eye.* Charles C Thomas, Springfield, Illinois

Franceschetti A (1963) *Ueber tapeto-retinale Degenerationen in Kindesalter. In: Entwicklung und Fortschritt in der Augenheilkunde.* Enke Verlag, Stuttgart, pp. 107–120

Krill AE, Deutman AF (1972) The various categories of juvenile macular degeneration. Trans Am Ophthalmol Soc 70: 220–245

Obituary (1927) Kl Mbl Augenheilk 78: 698–699

Stargardt K (1909) Über familiäre, progressive Degeneration in der Makulagegend des Auges. Graefe Arch Klin Exp Ophthalmol 71: 534–549

Stargardt K (1917) Ueber familiäre Degeneration in der Makulagegend des Auges, mit und ohne psychische Storungen. Arch Psychiat Nervenkr 58: 852–857

Wright RE (1935) Familial macular degeneration. Br J Ophthalmol 19: 160–165

# STEIN, Irving F.

*(1887–1976)*

*Courtesy*: Dr I.F. Stein Jr, Illinois

S TEIN-LEVENTHAL syndrome comprises hirsuitism, obesity, amenorrhea and polycystic ovaries. The condition is probably heterogeneous and in some families it is transmitted as a sex-limited autosomal dominant trait.

# Biography

S TEIN was a distinguished gynaecologist in Chicago, USA, during the first half of the present century.

Irving F. Stein was born on 19 September 1887 in Chicago, Illinois, USA. He qualified in medicine at the Rush Medical College in 1912 and trained in obstetrics and gynaecology at the Michael Reese Hospital. He was appointed to the attending staff in 1916 and retained this association until his death in 1976. Concurrently he held a senior academic appointment in obstetrics and gynaecology at the Northwestern University Medical School and he was also a senior member of the staff of the Highland Park Hospital.

Stein had an active academic career and he was a pioneer in the investigation of female infertility. His contributions were recognised by honorary membership of the medical societies of Venezuela, Peru, Brazil and Mexico and he became president of the International Fertility Association and the American Society for the Study of Sterility. During his long career Stein published 116 articles and 16 chapters in medical texts.

In addition to his academic achievements, Stein had a reputation as a warm, caring clinician and an excellent teacher. In his obituary, his friend and colleague, Dr MR Cohen, said of him: "Dr Stein was part of that grand era of Michael Reese Hospital when attending physicians, elegantly groomed, wearing a boutonniere, made rounds with a retinue of staff and nurses following. He was dignified and meticulous in everything that he did, whether it was patient care, surgery, or his own personal appearance. He was an excellent teacher and lecturer and an inspiration to all of us, especially the junior staff and students."

Stein's first wife, Lucile Oberfelder, died in 1940 and in 1954 he married Ruth W. Steit. He died in 1976, at the age of 89 years. His son, Irving F. Stein, jnr. became professor of surgery at the Northwestern University Medical school, Chicago.

# Nomenclature

S TEIN published extensively in many areas of obstetrics and gynaecology. His interests became increasingly focussed on infertility and in 1935, in collaboration with his colleague Michael Leventhal[1], he co-authored an article entitled *Amenorrhea associated with bilateral polycystic ovaries*. In this seminal work, which comprised reports of 7 cases and an extensive discussion, they were the first to document the association of defective ovulation, secondary amenorrhea and multiple cysts in both ovaries, in the presence of normal levels of follicle-stimulating hormone and 17-ketosteroids. In addition to delineating this disorder, Stein and Leventhal were able to demonstrate that wedge resection of the affected ovaries was an effective mode of treatment.

The importance of Stein and Leventhal's work was quickly recognised and the conjoined eponym came into general use. Attention was drawn to the genetic background of the disorder in 1968, when Cooper et al. showed that fathers and sisters of affected females were often unduly hirsute and that polycystic ovaries occurred with increased frequency in female siblings. In 1971 Givens et al. published a report of 2 large families in which 41 women had varying combinations of hirsuitism and oligomenorrhea, terming the condition "familial ovarian dysgenesis". Minor endocrine abnormalities and hirsuitism were present in some males in these families. Apparent transmission through a father and son had occurred and autosomal dominant inheritance was postulated. The genetic nature of the condition was further emphasised in 1972 by Borghi et al. in an article concerning affected sisters, in which the eponym Stein-Leventhal was again employed. Many articles followed, mostly concerning the underlying pathogenetic mechanisms. The Stein-Leventhal syndrome now represents a well documented and clinically important disorder.

# References

Borghi A, Maiello M, Giusti G (1972) Stein-Leventhal syndrome in sisters. The possible role of genetic factors in the "polycystic ovary syndrome". Acta Genet Med Gemellol 21: 79–93
Cooper HE, Spellacy WN, Prem KA, Cohen WD (1968) Hereditary factors in the Stein-Leventhal syndrome. Am J Obstet Gynec 100: 371–387
Givens JR, Wiser WL, Coleman SA, Wilroy RS, Andersen RN, Fish SA, Watson BS (1971) Familial ovarian hyperthecosis: a study of two families. Am J Obstet Gynec 110: 959–972.
Stein IF, Leventhal ML (1935) Amenorrhea associated with bilateral polycystic ovaries. Am J Obstet Gynec 29: 181–191

---

[1]Michael Leo Leventhal (1901–1971) was clinical professor of obstetrics at the University of Chicago.

# STRANDBERG, James Victor
*(1883–1942)*

*From*: Acta dermato-venereologica (1941) 22: 573
*Courtesy*: Scandinavian University Press, Stockholm

G ROENBLAD-STRANDBERG disease or pseudoxanthoma elasticum (*see* Groenblad, p. 73).

# Biography

S TRANDBERG occupied the prestigious chair of dermatology and syphilology at the Karolinska Institute, Stockholm, prior to World War II.

James Victor Strandberg was born in Stockholm on 9 March 1883. He was educated in that city and obtained his medical qualification in 1909. In 1912, after housemanship at the Sabbatsberger hospital, Strandberg entered private practice. He was a successful practitioner and was trusted by his patients and frequently consulted by his colleagues. His reputation was enhanced when he was the first to diagnose smallpox during the great epidemic of 1920.

Strandberg retained his academic associations and became a registrar at the Karolinska Institute in 1922, working with Professor Almkvist at the St Goran Hospital. He was appointed as associate professor in 1931 and in 1936 he was promoted to the post of full professor of dermatology and syphilology.

As an exceptional diagnostician and teacher he was held in high regard. He had a special interest in the treatment of venereal diseases and the ethical implications of these conditions. Strandberg's research interests included occupational eczema, sun damage and skin tuberculosis and he published many papers concerning the wider aspects of dermatology. In the later stages of his career he became increasingly involved in socio-medical matters. He was known for his tolerance and humanitarianism and because of these qualities, his assistance was frequently solicited by medical associations and civil authorities.

Strandberg was a friendly, hospitable man, who enjoyed entertaining in his home. He was happily married and had 2 sons, a daughter and several grandchildren. His main interests outside his professional life were sailing and fishing. Towards the end of his career Strandberg developed chronic ill-health due to hypertension. He died in his home in 1942 from a cerebral haemorrhage.

# Nomenclature

T HE major manifestations of Groenblad-Strandberg disease, or pseudoxanthoma elasticum (PXE), are in the eye and skin. A Swedish ophthalmologist, Ester Groenblad documented the ophthalmological features in 1929, while Strandberg gave a definitive description of the dermatological changes in the same year. These skin changes had in fact been documented in detail by Chauffard in 1889 in the following well known case report: "This is a man of 35 years.... At 24 years of age, while doing his military service in New Caledonia, he suffered a large hematemesis. This accident repeated itself several times (at age 26, 31, and 33). This summer he was admitted to the Hôtel Dieu for a hematemesis.

In 1880 L was discharged from the army and returned to France; and it was shortly afterward, he states, when he noted the beginning of his skin affection.... The xanthomatous eruption is composed of a series of evolving groups, perfectly symmetrical and confined exclusively to several flexural folds – the base of the neck, the two axillary creases, the folds of the elbows, the anterior abdominal wall, especially just below the umbilicus, the two inguinal triangles, the inferior aspect of the penis, around the anus, the two popliteal fossae.... The center of the group is formed by an almost confluent agglomeration of intradermal plaques, soft to touch, projecting to some extent like papules, separated by small folds of skin. Their coloration is rather pale, resembling that of fresh butter or yellow chamois; the size of the largest plaque is scarcely greater that that of a pea.... If one retracts both lips, one sees that the mucosa of the inner aspect is involved. It demonstrates a cluster of small, yellowish intramucosal nodules, resting on a richly vascular background traversed by numerous dilated and tortuous capillaries."

The conjoined eponym was employed for the disorder in the years following the publication of the articles by Groenblad and Strandberg in 1929. The descriptive term "pseudoxanthoma elasticum" in its shortened form "PXE" is now in general use.

# References

Chauffard A (1889) Xanthélasma disséminé et symétrique ans insuffisance hépatique. Bull Soc Méd Hop Paris 6 (Series 3): 412–419

Groenblad E (1929) Angioid streaks – Pseudoxanthoma elasticum: vorlaüfige mitteilung. Acta Ophthalmol 7: 329–336

Groenblad E (1932) Angioid streaks – Pseudoxanthoma elasticum. Der Zusammenhang Zwischen Diesen Gleichzeitig Auftredenden Augen – und Hautveränderungen. Acta Ophthalmol, Supp I, Vol 10

Hellerström S (1941) Personalia. James Strandberg. Acta Dermato-Venerol 22: 573–577

Strandberg J (1929) Pseudoxanthoma elasticum. Z Haut Geshlechtske 31: 689–693

# SULZBERGER, Marion B.

*(1895–1983)*

*From*: Shelley W.B., Crissey J.T. (1953) Classics in Clinical
Dermatology.

*Courtesy*: Charles C. Thomas, Springfield, Illinois
Portrait by Dorothy Wilding, New York

B LOCH-SULZBERGER syndrome, or incontinentia pigmenti (*see* Bloch, p. 29).

# Biography

S ULZBERGER was an outstanding academic American dermatologist. He lived a long, active and interesting life and made many contributions to his speciality.

Marion Baldur Sulzberger was born in New York City on 12 March 1895, the eldest son of Ferdinand and Stella Sulzberger. His father had emigrated to America from Bavaria in 1863 and had risen from the status of butcher boy to become the owner of one of the largest international meat packing firms in the world. He married 3 times and his son, Marion, was the offspring of the third union.

During his school years Marion Sulzberger was a brilliant scholar but nevertheless encountered a succession of problems. Refusing to go to college, he requested his father to take him into the family business. After a year of enthusiastic participation he lost interest and, although still only 16 years of age, took to an uninhibited lifestyle. This phase passed and he successfully applied for admission to Harvard but returned to his old roistering habits and left prematurely. During the next 2 years Sulzberger travelled to various parts of the world, working as a kitchen hand in Switzerland, a docker in England and a shepherd in Australia. Upon his return to America he learned that his father had died while he was overseas. With the impending involvement of the USA in World War I, Sulzberger decided to join the armed forces. He became a Naval aviator and, when his natural teaching abilities were recognised, he was promoted to the rank of flying instructor. During World War II Sulzberger rejoined the Naval Reserve as a lieutenant commander. He was subsequently decorated by the United States and France for his outstanding contributions to the understanding and treatment of the dermatoses caused by poison gases, burns and tropical skin diseases.

Sulzberger's decision to study medicine was only made when he left the Navy after World War I. He entered the medical school in Geneva in 1920, later transferring to the University of Zurich and within a short period his high academic abilities were recognised. It was during this period that the young Sulzberger came into contact with Professor Bruno Bloch and Professor Joseph Jadasshon, who were instrumental in fostering his interest in dermatology.

In 1929, having been well trained in European dermatology, Sulzberger returned permanently to America, where he entered private practice with Fred Wise, who was to become his close friend and mentor. He concurrently held a post in the New York Skin and Cancer Unit, ultimately becoming director in 1949 following the retirement of Dr George Miller MacKee. In the same year the Unit was incorporated into the department of dermatology and syphilology of the New York University-Bellevue Medical Center, with Marion Sulzberger as professor and chairman. Under his direction the department became the mecca of American dermatologists; he was acclaimed for his outstanding teaching abilities, infectious enthusiasm, insatiable curiosity and wide knowledge. He wrote numerous monographs and papers on many aspects of dermatology, founded the *Journal of Investigative Dermatology* and the Society of Investigative Dermatology and served on the editorial boards of several other journals. Sulzberger was internationally recognised and held honorary membership of many dermatology societies, including those of Great Britain, France and the Netherlands.

In 1961 at the age of 65 years Sulzberger retired from the chair of dermatology, but continued active associations with other institutions, universities, the armed forces and the public health service. His retirement was short lived, however, and after 3 years he was appointed as professor of clinical dermatology at the University of California in San Francisco. Thereafter, he was instrumental in the establishment of the Letterman Army Institute in San Francisco and served it as first director.

On 30 June 1970 Sulzberger retired once again but continued to teach at the University of California and to consult at the Letterman Army Institute. In collaboration with his wife, who was a writer and motion picture producer, Sulzberger established the Institute for Dermatologic Communication and Education, a non-profit organisation which made and distributed audiovisual aids and motion pictures on dermatological topics. Sulzberger died suddenly from a heart attack on 23 November 1983.

# Nomenclature

I N 1928, Sulzberger wrote an article in the German language, on the subject of "congenital pigmentary anomaly". Two years previously Bruno Bloch of Zurich had published an account of "incontinentia pigmenti", also in German, and thereafter the disorder bore their conjoined eponym (*see* Bloch, p. 29). An extensive review of the Bloch-Sulzberger syndrome has been provided by Landy and Donnai (1993).

Following delineation of the clinical and dermopathological manifestations, attention was focussed on the genetics of the disorder. The majority of affected persons are female and it seems probable that the condition is an X-linked dominant trait, with male lethality. In 1989 Sefiani and colleagues succeeded in assigning the locus of the faulty gene to band 28 of the long arm of the X-chromosome.

# References


Bloch B (1926) Eigentumliche, bisher nicht beschriebene Pigmentaffektion (incontinentia pigmenti). Schweiz Med Wochenschr 7: 404–405

Hunter JAA, Holubar K (1984) Sulzberger! Biography, Autobiography, Iconography. A posthumous festschrift. Am J Dermatopath 6/4: 345–370

Landy SJ, Donnai D (1993) Incontinentia pigmenti (Bloch-Sulzberger syndrome). J Med Genet 30: 53–59

MacKee GM (1955) Dr Marion B Sulzberger. J Invest Derm 24: 141–154

Sefiani A, Abel L, Heuertz S et al. (1989) The gene for incontinentia pigmenti is assigned to Xq28. Genomics 4: 427–229

Sulzberger MB (1928) Ueber eine bisher nicht beschriebene congenitale Pigmentanomalie (incontinentia pigmenti). Arch Derm Syph (Berl) 154: 19–32

# TARUI, Seiichiro
*(b. 1927)*

T ARUI disease (glycogen storage disease VII) is caused by hereditary deficiency of muscle phosphofructokinase (PFK) activity and leads to exercise intolerance, muscle cramps, increased haemolysis and myogenic hyperuricaemia. Inheritance is autosomal recessive.

# Biography

T ARUI, a Japanese physician, was professor of internal medicine at Osaka University of Foreign Studies in the middle decades of the twentieth century. He undertook pioneering studies of inherited glycolytic defects in human skeletal muscle.

Seiichiro Tarui was born in 1927 in Koshiën, Hyogo prefecture, Japan. During his childhood his family moved to Itami in the same prefecture, where his father owned an aluminium factory. After initial studies in the Faculty of Engineering at Kyoto University, Tarui qualified in medicine in Osaka in 1953, receiving the Kusumoto Prize. Postgraduate studies followed and Tarui completed a doctoral thesis on zinc metabolism in diabetes. In 1959 he married Yoshiko Iwasaki, a daughter of a Professor of Foreign studies at Osaka University and in the same year he became an assistant in the Second Department of Internal Medicine, Osaka University Medical School, where he continued his clinical research on disorders of carbohydrate metabolism. In 1966 Tarui was appointed as Associate Professor of Internal Medicine and in 1978 he was elevated to the Professorship and Chairmanship of Internal Medicine; he occupied this post until he retired with Emeritus status in 1991.

Although Tarui was a clinician, he always emphasised the importance of the exchange of knowledge between basic biochemistry and clinical medicine. He made important contributions as a diabetologist and lipidologist and his papers on preventive and therapeutic effects of nicotinamide on model mice with insulin-dependent diabetes mellitus attracted attention. Autoimmune hyperchylomicronemia was another unique clinical entity delineated by his group.

Tarui was active in many medical organisations and he was known for his warm humour and sympathetic manner. His contributions were recognised by several academic awards including the Hagedorn Prize (1990), Japanese Medical Association Medical Prize (1990), Uehara Prize (1991), and Takeda Medical Prize (1995).

After his retirement from the University, Tarui became Director of Otemae Hospital (Federation of National Public Service) near Osaka Castle. He also continued his academic activities and wrote several papers and essays including a description in 1995 of a novel variant of muscle PFK deficiency. In the summer of 1994, Tarui organised and chaired the International Symposium on Glycolytic and Mitochondrial Defects in Muscle and Nerve in Suita, Osaka.

Tarui wrote and spoke publicly on a wide variety of topics, including art and literature. He was also interested in philosophy and he once debated with a Japanese philosopher on the subject of "savoir de la clinique". In 1996 Tarui was enjoying a quiet life, taking pleasure in reading, golf, music concerts and playing the traditional "go game".

# Nomenclature

I N 1964 Tarui investigated 5 siblings, 3 of whom complained of rapidly induced fatigue and inability to keep pace with their classmates. His studies revealed that they were suffering from muscle PFK deficiency and in the following year he published a paper describing his findings under the title, *Phosphofructokinase deficiency in skeletal muscle. A new type of glycogenosis.* Thereafter, in 1968, Brown and Brown classified this entity as Type VII glycogenosis.

The eponym "Tarui" appeared in the *Dictionary of Medical Syndromes* (Magalini et al. 1990) and it was subsequently employed in textbooks by several authors, including Serratrice (1982) and Walton (1983). The term "Tarui disease (Type VII glycogenesis)" now enjoys wide acceptance.

The "myogenic hyperuricemia", which Tarui and his associates discovered is now recognised as an overproduction type of hyperuricaemia, where cell energy crisis causes increased degradation of ATP, resulting in elevated serum oxypurines and uric acid. In 1990 they finally succeeded in identifying the causal genetic defect of this disease; this represented the first demonstration of causal genetic defect among 8 types of glycogen storage disease.

Tarui disease shares several clinical manifestations with McArdle disease, a condition delineated by McArdle in 1951 (*see The Man Behind the Syndrome* p. 219). Increased haemolysis is completely lacking in McArdle disease, however, while myogenic hyperuricaemia is usually far milder in McArdle disease than in Tarui disease.

# References

Brown BI, Brown DH (1968) Glycogen-storage diseases: Types I, III, IV, V, VII and unclassified glycogenoses. In: Dickens F, Randle PJ, Whelan WJ (eds) Carbohydrate Metabolism and its Disorders, Vol 2 p. 123–150. Academic Press, London & New York

Magalini SI, Magalini SC, Francisci G (1990) Tarui's Syndrome. In: Dictionary of Medical Syndromes, p. 860. JB Lippincott Co, Philadelphia

McArdle B (1951) Myopathy due to a defect in muscle glycogen breakdown. Clin Sci 10: 13–35

Nakajima H, Kono N, Yamasaki T, Hotta K, Kuwajima M, Noguchi T, Tanaka T, Tarui S (1990) Genetic defect in muscle phosphofructokinase deficiency. Abnormal splicing of the muscle phosphofructokinase gene due to a point mutation at the 5'-splice site. J Biol Chem 265: 9392–9395

Serratrice G, Gastaut JL, Pellissier JF, Pouget J, Desnuelle C, Cros D (1982). Deficit en phosphofruktokinase musculaire (glycogenose type VII ou maladie de Tarui). In: *Maladies Musculaires*, Masson, Paris, pp 191–192

Tarui S (1995) Glycolytic defects in muscle: aspects of collaboration between basic science and clinical medicine. Muscle and Nerve, Suppl 3: S2-S9

Tarui S, Okuno G, Ikura Y, Tanaka T, Suda M, Nishikawa M (1965) Phosphofructokinase deficiency in skeletal muscle. A new type of glycogenosis. Biochem Biophys Res Commun 19: 517–523

Tarui S, Kono N, Nasu T, Nishikawa M (1969) Enzymatic basis for the coexistence of myopathy and hemolytic disease in inherited muscle phosphofructokinase deficiency. Biochem Biophys Res Commun 34: 77–83

Walton JN (1983) Tarui's disease. In: Weatherall DJ, Ledingham JGG, Warrell DA (eds) *Oxford Textbook of Medicine*. Oxford University Press, Oxford, pp. 22.20 only

# TOURAINE Albert

## *(1883–1961)*

*From*: Brit J Dermatol (1961) 73: 346 (Obituary)

*Courtesy*: Blackwell Science Ltd, Oxford

T OURAINE syndrome or centrofacial neurodysraphic lentiginosis is a familial, possibly autosomal dominant disorder in which facial freckles are variously associated with mental retardation.

# Biography

T OURAINE was a leading French dermatologist during the first half of the twentieth century.

Albert Touraine was born in Paris on 11 November 1883. He was educated at the Lycée Charlemagne and studied medicine at the University of Paris. He was an outstanding student and received a gold medal in 1911, having studied under such luminaries as Apert, Netter and Achard. Touraine developed an early interest in venereology and his thesis submitted in 1912 contained original observations of agglutination reactions in syphilis.

In 1914, Touraine was mobilised for war service with the 166th Infantry Regiment. He spent two years in the Verdun region and was then promoted Chief of the Dermatology and Syphilology service at a specialist hospital in Besanÿon. His contributions were rewarded by his appointment as Chevalier of the Legion of Honour and in later life he was advanced to the status of Officer and finally Commander. After the armistice, Touraine held successive appointments at La Charité and Broca hospitals, Paris and in 1932 the final step in his career took him to the senior post of physician to the St Louis Hospital, where he remained until he died in 1961.

Touraine had immense energy and in addition to his clinical activities, he was a prodigious medical author. During his professional lifetime, more than 1200 medical articles and reviews flowed from his fluent pen. Touraine also found the time to edit the journal *Annales de Dermatologie et de Syphiligraphie* for the last 18 years of his life.

Touraine played a major role in a number of medical organisations and he also served as Vice-President of the French Society of Genetics, and as President of the French Society of Dermatology and Syphilology. In 1960 he received the accolade of election to the Vice Presidency of the Academy of Medicine.

In his prime, Touraine was a tall man, with blue eyes and a pointed beard. His personal qualities were summed up in his obituary: "He was naturally respected for his immense industry and erudition and for his intellectual integrity, but he perhaps won even more admiration for his natural humour and warm-hearted friendliness. Those who became better acquainted with him learned that he was what he appeared to be, a conservative Frenchman, devoted to his family and deeply attached to the traditions and well tried institutions of his country."

Touraine became unwell in 1960 but he continued his editorial duties until the day of his death in May 1961.

# Nomenclature

I N 1941, Touraine published an account of 17 families, comprising 32 affected persons, in which lentigines of the central portion of the face were associated, in some instances, with mental retardation. In 8 of the families, parent to child transmission had occurred, while in 5 families, pairs of siblings were affected.

Touraine gave further details of the disorder in his massive text *Heredity in Medicine* which was published in 1955. Although the condition is uncommon, Dociu and colleagues (1976) were able to identify 40 cases in the literature. Irregular autosomal dominant transmission seems likely, although this remains unproven.

In addition to the lentiginous disorder, Touraine's name is associated with the nail-patella and Christ-Siemens-Touraine syndromes. The term "Touraine aphthosis" is also used as a synonym for the form of non-genetic oral ulceration which conventionally bears Behçet's eponym.

# References


Dociu I, Galaction-Nitelea O, Sirjita N, Murgu V (1976) Centrofacial lentiginosis. Brit J Derm 94: 39–43

Duperrat B, Golé L (1962) Nécrologie. Albert Touraine 1883–1961. La Presse Médicale 70(3): 153–154

Obituary. Dr Albert Touraine (1961). Br J Derm 73: 346–347

Ronchese F (1962) Obituary. Albert Touraine, MD 1883–1961. Arch Derm 85: 85–286

Touraine A (1941) Une nouvelle neuro-ectodermose congenitale: la lentiginose centro-faciale et ses dysplasies associées. Ann Derm Syph 8: 453–473

Touraine A (1955) *L'Heredité en Médecine*. Masson, Paris

# ULLRICH, Otto

## *(1894–1957)*

*Courtesy*: Professor H.-R. Wiedemann, Kiel

U LLRICH-TURNER syndrome comprises stunted stature, poorly developed secondary sexual characteristics, a typical facies and variable cardiac malformations. Affected persons, who are all female, have an aberrant or absent X-chromosome.

# Biography

U LLRICH was a German paediatrician and a founder of clinical genetics in his country. He made many contributions to syndromic delineation and dysmorphology.

Otto Ullrich was born in Saxony in 1894 and became a medical student in Munich. During World War I he served as an assistant in the medical corps and after the armistice in 1918 he worked with Professor Meinhard von Pfaundler, the chairman of paediatrics at the University of Munich. Pfaundler was a firm believer in the importance of genetics in paediatrics and he had a profound influence on the development of that discipline and upon Ullrich's future academic direction. In 1922 Ullrich was appointed to the directorship of the Policlinic and by 1929 he had achieved faculty status. Ullrich moved to Berlin in 1934, as director of the National Centre to Combat Infant Mortality and it is of interest that a knowledge of human genetics was a prerequisite for this post. Ullrich was unhappy with the political atmosphere in Berlin, and after 6 months he moved to Essen, as director of the Municipal Children's Hospital. In 1939 Ullrich was called to the chair of paediatrics at Rostock and in 1943 he took up the chair in Bonn, where he remained until his death in 1957.

Ullrich made major contributions in several scientific and medical fields, including biochemistry, metabolic disorders, haematology, neurology and dermatology. He had a special interest in genetic conditions and, in addition to those which bear his eponym (vide infra) he published significant articles on many topics, including the mucopolysaccharidoses, endocardial fibroelastosis, choanal atresia and renal rickets. In the broader context, Ullrich developed concepts concerning multiple congenital abnormalities, phenotypic variability, pleiotropy and heterogeneity, all of which are fundamental to contemporary medical genetics.

Professor H.-R. Wiedemann, Professor Emeritus of the University of Kiel and Ullrich's former chief resident, gave this account of his mentor's professional and personal attributes: "As physician, Ullrich was very broadly based. At the bedside he was patient and was quickly able to win the trust of the child. He examined very calmly and was an excellent observer with a capacity to take in the essentials of the case at a glance. The laboratory data were reviewed routinely but were never overemphasized, quite in the highly critical and analytical Munich spirit. His special ability to retain and to recall previous cases allowed Ullrich to make correct diagnoses with surprising ease. His rounds were as punctual as clockwork and very thorough, offering a wealth of information and experienced counsel, enriched by his critical perspective and pronounced distaste against hastiness, especially in therapy. This was not only a result of his Munich background but also reflected his personal inclination.

As a human being Otto Ullrich had a gracious and noble personality, with a compelling glance and a care for moderation and compromise. Apart from professional contacts, Ullrich could be reserved with younger co-workers; however, when he was able to open himself to others he always engendered much joy."

Ullrich's achievements were honoured in 1991 by the establishment of the Otto Ullrich medal for excellence in medical genetics. The creation of this award was announced in the *American Journal of Medical Genetics*, in an edition that contained an editorial and several articles pertaining to Ullrich and his scientific achievements.

# Nomenclature

I N 1929 at a meeting of the Munich Paediatric Society, Ullrich presented a girl aged 8 years with abnormalities which included a webbed neck, stunted stature, cubitus valgus and an unusual facial appearance. He published an account of this patient in the following year, in the German literature.

In 1938 Henry Turner (*see The Man Behind the Syndrome, p. 175*) reported 7 young women with sexual infantilism, laxity of the skin of the neck, short stature and retarded bone age, all of which he attributed to a primary defect in the anterior pituitary gland. The autonomous status of the syndrome was confirmed in 1959 when Ford et al. demonstrated that affected females lacked an X-chromosome. Thereafter the condition was generally known as the Turner syndrome in the American and English literature, and as the Ullrich-Turner syndrome in publications from the Continent of Europe.

At the time that Turner was studying his patients in the USA, Kristine Bonnevie (*see p. 33*) was undertaking animal experiments in Germany. Her investigations led to concepts concerning human malformation, with which Ullrich became involved, and the eponym "Bonnevie-Ullrich" syndrome was applied to males with the Turner or Ullrich-Turner syndrome phenotype, or to females with the phenotype but a normal chromosomal constitution. The eponym "Noonan" (*see p. 121*) is currently applied to these persons, and the name of Bonnevie is now little used. Nosological difficulties are compounded by the fact that a woman aged 25 years with the classical Ullrich-Turner phenotype was documented in 1925 by Seresevskij; this eponym is preferred by Russian colleagues.

Until recently there has been controversy as to whether Ullrich's contribution warranted eponymous recognition, as some authors believed that the girl whom he had described in 1930 actually had the Noonan syndrome. This issue was settled in magnificent style in 1991, when Wiedemann and Glazl restudied this patient, now 66 years of age, and demonstrated that her chromosome constitution was 45, XO. In their article these authors depicted the patient in early childhood and mature adulthood and advanced a forceful demand for the recognition of historical priority and the retention of the conjoined eponym Ullrich-Turner.

# References


Ford CE, Jones KW, Polani PE, de Almeida JC, Briggs JH (1959) A sex-chromosome anomaly in a case of gonadal dysgenesis (Turner's Syndrome). Lancet I: 711

Opitz JM (1991) Editorial Comment. Otto Ullrich: An appreciation. Am J Med Genet 41: 126–127

Seresevskij NA (1925) In relation to the question of a connection between congenital abnormalities and endocrinopathies. The Russian Endocrinological Society on 12th November

Turner HH (1938) A syndrome of infantilism, congenital webbed neck and cubitus valgus. Endocrinology 23: 566–574

Ullrich O (1930) Über typische Kombinationsbilder multipler Abartungen. Z Kinderheilk 49: 271–276

Ullrich O (1949) Turner's syndrome and status Bonnevie-Ullrich. Am J Hum Genet 1: 179–202

Wiedemann H-R (1991) Otto Ullrich and His Syndromes. Am J Med Genet 41: 128–133

Wiedemann H-R, Glazl CJ (1991) Follow-up of Ullrich's original patient with "Ullrich-Turner" syndrome. Am J Med Genet 41: 134–136

# VAN BOGAERT, Ludo

*(1897–1989)*

*Courtesy:* Professor D. Klein, Switzerland

V AN BOGAERT-Scherer-Epstein syndrome, or cere-brotendinous xanthomatosis is a slowly progressive autosomal recessive disorder in which cholesterol accumulates in the brain, lungs and tendons.

# Biography

V AN BOGAERT was a Belgian academic neurologist of the twentieth century. His prodigious research output is reflected in the fact that his name is attached to more than 10 separate disorders.

Ludo van Bogaert was born on 25 May 1897 in Antwerp, Belgium, where his father was a medical practitioner. He received his schooling at the Notre-Dame College in his home city and in 1914, at the onset of World War I, he left occupied Belgium for the University of Utrecht, where he commenced medical studies. When he reached the age of 18 years van Bogaert volunteered for the Belgian army, joining via England. He was twice wounded and sustained spinal injuries but recovered in time to accept a commission and serve with the occupation forces on the Rhine.

Van Bogaert resumed medical studies at the Free University of Brussels, graduating with distinction in 1922. His spinal wound had prompted an interest in neurology and he spent a year in Paris working at the Salpêtrière under Pierre Marie, and at the Hôpital de la Charité under Marcel Lebbe, gaining experience in neurology and neuropathology. On his return to Antwerp, van Bogaert became assistant at the St Elizabeth Hospital and later staff physician at the Stuivenberg Hospital. In 1925 he obtained his doctorate with a thesis on amyotrophic lateral sclerosis.

During the decade that followed van Bogaert published numerous articles on neurological topics. His prodigious academic output secured his reputation and in 1933 he accepted an appointment as director of the clinical services and pathological laboratory of the newly established Bunge Research Institute. The Institute was highly productive but the research team which van Bogaert had gathered together were dispersed at the outbreak of World War II. The region around the Institute suffered from heavy bombing and it was necessary for van Bogaert to move his laboratory and collection of specimens to the cellars of the Stuivenberg Hospital, where he was able to continue with his investigations. During the two decades which followed World War II, additional wings were built onto the Bunge Institute, and available space was tripled. Postgraduates arrived from many parts of the world and a comprehensive research programme was undertaken. He received the ultimate recognition in 1957 when he was elected as president of the newly formed World Federation of Neurology.

Van Bogaert maintained academic links with neurological centres throughout the world, and he had special ties with his neighbouring colleagues in France and Germany. During World War II he maintained these scientific contacts, although he had remained firm in his patriotic principles. His relationships with the German neurologists, Hallervoden and Spatz, were the subject of controversy in the immediate post-war years but, in general, van Bogaert emerged with his reputation intact.

Philippart, who wrote van Bogaert's biography in *Founders of Child Neurology*, made the following comments on his personal traits and qualities: "Van Bogaert's personality defies sketches. He was the kind of person one meets only once in a lifetime. He was elemental like the sea, powerful and illusory. People and facts were taken and understood as they were. The intimate feelings of this master diplomat remained carefully hidden behind his open and kind manner. His activities were prodigious but under complete control. There was never any sense of inner or outer pressure. He never gave orders or proffered opinions. Everything and everyone fell into place, and the work was completed. As one would expect from a man whose scientific tool was the microscope, he was a visual person. Inconspicuous details of a patient's examination were grasped within minutes and sorted out against his well-furnished memory. Even the most humble, inarticulate patient was treated with compassion and provided with warm, reasonable reassurance.

No ivory tower hermit, he was eminently sociable. His extraordinarily busy professional life did not preclude innumerable extracurricular activities. The range of his cultural and other interests was vast and varied, encompassing painting, literature and precious books, to cite but a few. He was a brilliant conversationalist, equally able to captivate the attention of colleagues, young assistants, or people of social renown and importance."

Van Bogaert died in Antwerp, Belgium on 4 March 1989, at the age of 92 years.

# Nomenclature

I N 1937 van Bogaert, together with his colleagues Scherer and Emile Epstein, a biochemist of Vienna, investigated 2 affected cousins and documented a progressive neurodegenerative disorder which was characterised by accumulation of cholesterol in the white matter of the brain. Xanthoma of the tendons and lungs were also present but serum cholesterol levels were within normal limits. Onset is in mid-childhood and progression is slow; in 1969 van Bogaert was able to publish a 30-year follow-up of one of his original patients. There have been numerous reports of this disorder and autosomal recessive inheritance is well documented. The title "cerebrotendinous xanthomatosis" is in general use but the conjoined eponym "Bogaert-Scherer-Epstein" is occasionally employed.

Van Bogaert's name is linked to that of Divry in the eponymous title of diffuse cortical angiomatosis, which they documented in 3 affected brothers in 1946 (Paul Divry, born 1898, was a Belgian physician). This autosomal recessive disorder presents in the second decade with progressive epilepsy, mental retardation and a diffuse mottling of the skin due to widespread telangiectasia. Pyramidal and extrapyramidal signs can be elicited and at autopsy diffuse angiomatosis is present in the meninges.

In addition to the above mentioned genetic disorders, van Bogaert's name is also applied to subacute sclerosing panencephalitis. He is also linked eponymously with several colleagues, including those of Canavan, Bertrand, Nijssen, Greenfield and Peiffer in the titles of a number of rare neurological disorders of uncertain aetiology.

# References


Karcher D (1990) In memoriam: Ludo van Bogaert (1897–1989) J Neuropath Exp Neurol 49: 185–187

Philippart M, van Bogaert L (1969) Cholestanolosis (Cerebrotendinous Xanthomatosis). A follow-up study on the original family. Arch Neur 21: 603–610

Philippart M (1990) Ludo van Bogaert. In: Ashwal S (ed) *Founders of Child Neurology*., Norman Publishing, San Francisco, pp. 854–861

Van Bogaert L, Scherer HJ, Epstein E (1937) *Une forme cérébrale de cholestérinose généralisée (type particulier de lipidose à cholestérine)*. Paris, Masson.

Van Bogaert L, Divry P (1945) Sur une maladie familiale caracterisee par une angiomatose diffuse cortico-méningée et une démyélinisation de la substance blanche du centre ovale. Bruxelles Med 25: 1090–1091

# VAN DER HOEVE, Jan

## *(1878–1952)*

*From*: Br J Ophthalmol (1952) 36: 399
*Courtesy*: BMJ Publishing Group, London

 AN DER HOEVE syndrome comprises brittle bones, blue sclerae and deafness. The condition is now generally known as osteogenesis imperfecta tarda, or OI type I.

# Biography

V AN DER HOEVE was a prominent Dutch ophthalmologist during the early decades of the twentieth century.

Jan van der .Hoeve was born in Santpoort near Velsen, Holland, on 13 April 1878. He graduated at the University of Leyden and obtained a doctorate at the University of Bern in 1902 with a thesis on ocular movements. From 1905 to 1913 van der Hoeve served as a military medical officer and on leaving the army he was appointed as professor of ophthalmology at the University of Gröningen. In 1918 van der Hoeve moved to his alma mater at Leyden as professor of ophthalmology, where he spent the remainder of his career.

Van der Hoeve had wide clinical interests and made numerous academic contributions. In particular, he was the first to introduce the concept of the phakomatoses as a group of congenital disorders with ramifications in many tissues. At the height of his career van der Hoeve was the foremost ophthalmologist in Holland and he did much to influence the development of his speciality. He was a competent linguist and had a wide circle of friends and colleagues in the many countries which he had visited. Van der Hoeve was courteous and genial, with considerable erudition and organising ability. These attributes were used to good effect in the re-establishment of international links following World War I and van der Hoeve served as chairman when the first post-war International Congress of Ophthalmology was held in Holland. He received an honorary degree from the University of Edinburgh and the Sir William Mackenzie medal for ophthalmic research. His wide scientific knowledge also led to his election in 1932 as president of the Physical Section of the Royal Dutch Academy of Science.

In his later years van der Hoeve became frail but retained his warm personality. He died on 26 April 1952 in Leyden, at the age of 74 years, following a road traffic accident.

# Nomenclature

I N the evolution of the understanding of osteogenesis imperfecta, Lobstein (1835) described an affected adult in his anatomical textbook, while Vrolik (1849) depicted the skeleton of a neonate in an embryology text. Thereafter, the eponym of Lobstein was applied to the late-onset or tarda variety of OI, while that of Vrolik was linked to the congenital or infantile form.

The manifestations of the condition were further delineated in 1896, when John Spurway of Tring, England, re-emphasised the association of skeletal fragility with blue sclerae. In 1900 Alfred Eddowes of London addressed the cause of the scleral blueness and proposed that it might be due to a defect of mesenchyme, while in 1912 Charles Adair-Dighton of Liverpool drew attention to the dominant mode of transmission of blue sclerae and the association with adult-onset deafness. In 1918 van der Hoeve and his colleague, de Kleyn of Utrecht, emphasised the syndromic relationship of brittle bones, blue sclerae and deafness in OI tarda. Thereafter, the identity of this form of OI became firmly established, and the eponym of van der Hoeve was sometimes used as a title, although the name of his co-author, de Kleyn was not employed. Van der Hoeve's own eponym has now fallen into disuse.

Osteogenesis imperfecta is one of the most frequent of the heritable disorders of connective tissue. The recognition of heterogeneity and the detection of a defect in type I collagen has generated considerable scientific interest and more than 1000 articles concerning OI have been published.

(For further details of the evolution of the nomenclature, *see* Lobstein p. 109 and Vrolik p. 183.)

# References


Adair-Dighton CA (1912) Four generations of blue sclerotics. Ophthalmoscope 10: 188–189

Eddowes A (1900) Dark sclerotics and fragilitas ossium. Br Med J 2: 222

Ekman OJ (1788) *Dissertatio medica. Descriptionem et casus aliqot osteomalaciae sistens.* Uppsala, Sweden

Lobstein JG (1835) *Lehrbuch der pathologischen Anatomie.* Stuttgart, p. 179

Obituary (1952) Br J Ophthal 36: 399–400

Spurway J (1896) Hereditary tendency to fracture. Br Med J 2: 844

van der Hoeve J, de Kleyn A (1918) Blaue Sclerae, Knochenbruchigkeit und Schwerhorigkeit. Graefes Arch Ophth 95: 81–93

Vrolik W (1844–1849) Tabulae ad illustrandam embryogenesim horninis et mammalium, tam naturalem quam abnormem. Amstelodami

# VOGT, Alfred

*(1879–1943)*

V OGT congenital cerulean cataract has a characteristic blue colour and a frosted appearance. Visual impairment is not usually severe although lens extraction may be necessary in adulthood. Inheritance is autosomal dominant.

# Biography

V OGT was head of the Ophthalmological Institute at the University of Zurich for two decades during the first half of the twentieth century.

Alfred Vogt was born on 31 October 1879 in Menziken, Aargau, Switzerland, where his father was a farmer and school teacher. His scientific abilities were recognised at an early stage and he was encouraged to study medicine. After qualifying at the University of Basel he trained in ophthalmology under Professor Mellinger in that city, writing a treatise *Damage to the Eye caused by Aniline Dye*. Vogt commenced private practice in Aargau in 1906 and in 1909 he was appointed as chief of the ophthalmic department at the local hospital.

Vogt had continued with his academic activities while in private practice and in 1917 he became professor of ophthalmology at the University of Basel. In 1923 Vogt accepted an invitation to head the Ophthalmic Institute at the University of Zurich and he occupied this post with distinction for the next two decades. Vogt was a firm believer in the efficacity of hierarchical structures; a biographer stated "Vogt was the omnipotent ruler, venerated by the resident and nursing staff, second only to God. When he spoke, that was it, and nobody attempted to argue". He went on to say "The Zurich Eye Clinic under Professor Vogt certainly was an efficiently run, hardworking and productive institute. Vogt kept close reins on the staff and by strict discipline, which would make some of our contemporary house staff shudder, produced maximal performance".

Contributions to the science of ophthalmology made by Vogt included the development of techniques for retinoscopy and the surgical management of retinal detachment. His most important research commenced in 1913, when he used the newly constructed slit-lamp in conjunction with a corneal microscope to examine the structures of the anterior regions of the eye. With this instrument he documented and published the morphological changes in many disease processes, eventually bringing his findings together in a 3-volume *Textbook and Atlas of the Slit-Lamp Microscopy of the Living Eye*. The second edition of this classical work was published in 1940.

Vogt was recognised as a leading figure in ophthalmology and he received the prestigious Donders and Gullstrand medals for his achievements. In 1939 his sixtieth birthday was marked by a Festschrift, with 90 papers contributed by colleagues in all parts of the world.

Vogt was saddened by the death of his only son in an avalanche in the Swiss mountains and his career at Zurich was impeded by chronic ill health due to a renal disorder, with vascular complications. He died in 1943 at the age of 64 years.

# Nomenclature

V OGT became interested in genetics through his extensive investigation of familial factors in senile changes in the eyes. He approached this problem by studying 19 pairs of elderly identical twins whom he had contacted by placing advertisements in the newspapers.

In 1922, using the slit-lamp, Vogt documented the features of the heritable cataract which bears his name. The blueish colour of this lesion prompted the term "cerulean", which pertains to the sky, while the alternative descriptive name "frosted cataract" is derived from the appearance produced by concentric layers of opacities within the affected lens.

In addition to the familial cataract, Vogt's name is attached to several other ophthalmological conditions. These include the Vogt senile floury cornea, Vogt corneal dystrophy and the autosomal dominant Vogt mosaic corneal degeneration. This latter abnormality has an appearance which has been likened to that of crocodile skin.

In 1906 Vogt documented an unusual uveocutaneous syndrome and in 1929 Yoshizo Koyanagi, a Japanese physician writing in the German literature, described additional syndromic components. Their eponyms are now paired in the title of this rare disorder, which develops in adulthood and comprises uveitis, iritis and glaucoma, with vitiligo, alopecia and deafness. The eponym of Harada is sometimes applied to an acute form of the disorder, which is probably caused by viral infection.

# References


Koyanagi Y (1929) Dysakusis, Alopecie und Poliosis bei schwerer Uveitis nicht traumatischen Ursprungs. Klin Mbl Augenh 82: 194–211

Stocker FW (1944) Obituary. Professor Alfred Vogt (1879–1943) Arch Ophthalmol 31: 172–174

Schläpfer H, Wagner H (1979) Zum 100 Geburtstag von Prof Alfred Vogt. Schweiz Med Wschr 109: 1565–1566

Stocker FW (1974) Remembrances of Things Past. 1923–1943: The era of Alfred Vogt. Survey of Ophthalmology 19: 31–37

Vogt A (1906) Frühzeitiges Ergrauen der Zilien und Bemerkungen über den sogenannten plötzlichen Eintritt dieser Veränderung. Klin Mbl Augenh 44: 228–242

Vogt A (1922) Weitere Ergebnisse der Spaltlampenmikroskopie des vorderen Bulbusabschnittes. III Angeborene und früh erworbene Linsenveränderungen. Graefes Arch Ophth 107: 196–240

# VOGT, Heinrich

## *(1875–1957)*

S PIELMEYER-VOGT disease, or juvenile amaurotic familial idiocy, is an autosomal recessive, neuro-degenerative disorder with onset in early childhood and progressive visual and mental dysfunction.

# Biography

V OGT was a German neurologist and psychiatrist during the early years of the twentieth century. After making several significant contributions to neuropsychiatry, he changed direction in mid-career and became a medical hydrologist.

Heinrich Vogt[1] was born in Regensburg, where his father was a school teacher, and he was educated at the prestigious St Stephan Gymnasium, Augsburg, Bavaria. He studied medicine at Munich, Heidelberg and Göttingen, becoming interested in neurology at an early stage in his career and wrote his doctoral thesis at the University of Heidelberg on the subject of *Facial Nerve Palsy after Acute Otitis Media*.

After qualification, Vogt joined the staff of the State Hospital for the mentally retarded and epileptic at Langenhagen, near Göttingen. He was greatly influenced by Alzheimer's attempts at delineation of specific neurological syndromes on a basis of clinico-pathological correlations and his post at Langenhagen provided him with the opportunity to pursue investigations of this type, including documentation of the disorder which now bears his eponym. He also made the original observation of the triad of mental retardation, adenoma sebaceum and epilepsy as components of tuberous sclerosis.

Vogt became professor of psychiatry at Göttingen in 1907 and in 1909 he moved to the University of Frankfurt as director of the department of psychiatry at the Senckenberg Institute. He published extensively on brain malformations and the institutional management of persons with mental retardation and epilepsy, and he gained wide recognition as an expert in these fields. He was a proponent of special courts for juvenile offenders and for the humane management of the mentally handicapped. Vogt became director of a neurological sanatorium at Wiesbaden in 1911. His publications continued, including a book on epilepsy and chapters on genetic disorders of the brain and neurological damage due to war wounds.

In 1925, for reasons which never became known to his colleagues and friends, Vogt moved to a spa, Bad Pyrmont, where he set up practice as a medical hydrologist. He took an academic approach to this new career and subsequently founded a research institute of balneology at Breslau and edited a journal on the subject. At the end of World War II Breslau was incorporated into Poland and Vogt moved back to Bad Pyrmont, where he remained in private practice until his death in 1957.

# Nomenclature

I N the early years of the twentieth century, mental retardation was largely regarded as a homogeneous disorder for which custodial care was the main aspect of management. In this context the term "idiocy" had the connotation of progressive mental degeneration. While employed at the Langenhagen Institution, Vogt investigated a number of children with idiocy and amaurosis (i.e. visual problems) and he recognised a specific condition which was familial, with onset in the middle years of the first decade. He compared the natural history of this disorder with the familial amaurotic idiocy previously delineated by Warren Tay

(*see The Man Behind the Syndrome* p. 165) and Bernard Sachs (*see The Man Behind the Syndrome* p. 155), concluded that these conditions were separate entities and proposed subdivision into infantile and juvenile forms.

Following clinical delineation, Vogt undertook histological investigations in his patients and demonstrated the presence of inclusion bodies in neuronal tissue. He submitted an account of his findings for publication, but in November 1905, prior to the appearance of Vogt's paper, Walther Spielmeyer (*see* p. 155) presented similar observations at a medical congress in Karlsruhe. Thereafter the condition became known as Spielmeyer-Vogt disease, or juvenile amaurotic familial idiocy. The eponym of Batten has also been associated with this disorder (*see The Man Behind the Syndrome* p. 13). Later, when the biochemical defect had been elucidated, the term "neuronal ceroid lipofuscinosis type 3" came into use. The conjoined eponym has been retained and, although purists claim that the order of the names should be reversed because of historical priority, the current format is well established.

# References


Reich, Weindl (1990) In: Ashwal S (ed) *Founders of Child Neurology*. Norman Publishing, San Francisco, pp. 614–629

Spielmeyer W (1905) Über familiäre amaurotische Idiotien. Arch Psych Nervenkrankh 140: 1038–1039

Vogt H (1905) Über familiäre amaurotische Idiotie und verwandte Krankheiten. Mschr Psych Neur 18: 161–171, 310–357

Vogt H (1907) Zur Diagnostik der tuberösen Sklerose. Zeit Erforsch jugendl Schwachsinns 2: 1–15

Vogt H (1911) Familiäre amaurotische Idiotie, histologische und histopathologische Studien. Arch Kinderheilk 51: 1–25


---

[1]In addition to Heinrich Vogt, several other persons with this family name have gained eponymous immortality:

Alfred Vogt (1879–1943) was head of ophthalmology at the University of Zurich (*see* p. 175);

Cecille Vogt (1875–1962) was a French physician working in Germany. She published with O. Vogt on a progressive choreo-athetoid neurodegenerative disorder of infancy;

Oskar Vogt was director of the Neurobiological Institute of Berlin in the early decades of the present century (*see* Bielschowsky, p. 25).

# VON GRAEFE, Albrecht Friedrich Wilhelm

## (1828–1870)

*From*: Lebensohn J.E. (1969) An Anthology of Ophthalmic Classics
*Courtesy*: The Williams & Wilkins Co, Baltimore, Maryland

G RAEFE-SJOGREN syndrome is an autosomal recessive disorder comprising retinitis pigmentosa, congenital deafness and spinocerebellar ataxia.

# Biography

V ON Graefe was the pre-eminent ophthalmologist of his era and he is regarded as the founder of modern clinical ophthalmology.

Albrecht von Graefe was born on 8 May 1828 in Berlin, the third son of Carl Ferdinand von Graefe, a distinguished surgeon, who was director of the University Surgical Clinic. Von Graefe's family were wealthy and he had a happy boyhood. As a scholar he was clever, intelligent and diligent and he entered the University of Berlin at the age of 15 years. Von Graefe studied philosophy, logic, natural sciences and anatomy, and as his medical education proceeded, he became increasingly interested in ophthalmology. He obtained experience in Prague, Paris and Frankfurt before undertaking further studies in Vienna. After a visit to Moorfields Hospital in London, von Graefe settled into practice in Berlin, where he established a free ophthalmological service for indigent persons. The "Graefe Clinic" became famous and attracted large numbers of patients; by the time that he was 30 years of age von Graefe had treated or operated upon more than 10,000 individuals. Von Graefe had immense reserves of energy and in addition to his punishing work schedule, he was able to devote time to his physiological and other scientific studies.

In 1852 von Graefe moved his clinic to 46 Karlstrasse, Berlin, where a wall tablet still commemorates his occupancy. In the same year he submitted a thesis entitled *On the action of the ocular muscles* to the University of Berlin and received the status of "privatdocent". Von Graefe was an excellent teacher and postgraduates from all parts of the world attended his clinics. Weber of Darmstadt commented "One was spell-bound in his clinic, as if in a magic place. The multitude of new facts and viewpoints never heard before, the fascinating presentation and glowing enthusiasm acted like a revelation".

By 1854 the discovery of the ophthalmoscope by Helmholtz had revolutionised ophthalmology and von Graefe recognised the need for a journal to cater for the advances in the speciality. He duly launched *Archiv für Ophthalmologie* which he edited; it is remarkable than von Graefe was only 26 years of age at this time. In 1857 von Graefe was appointed as assistant professor at the University of Berlin. Shortly afterwards, at the International Congress of Ophthalmology at Brussels, his presentation of a proposal for the treatment of glaucoma by iridectomy received acclamation and secured his reputation. In the same year, together with friends in Heidelberg, he founded the German Ophthalmology Society. Von Graefe invented a special knife, which is still used for cataract surgery and he was the first to use Helmholtz's ophthalmoscope. Von Graefe is rightly regarded as the founder of modern clinical ophthalmology.

In personal appearance von Graefe was a tall, slender and elegant with a handsome face, long dark hair and a full beard. He was modest in his lifestyle and his greatest pleasure was entertaining his friends to dinner in his home. His massive workload, which he enjoyed, left time for little else!

In 1861 von Graefe developed tuberculous pleurisy but fortunately his condition went into remission. He married Anna Gräfin Knuth in 1862 and 5 children were born in the next few years. Von Graefe became the first German professor of ophthalmology in 1866 and after initial difficulties, became head of the Charity Clinic and established ophthalmology in the undergraduate curriculum. This new initiative put a strain upon his health, his pulmonary tuberculosis was reactivated and haemoptysis became troublesome. Von Graefe died on 20 July 1870 at the age of 42 years.

# Nomenclature

I N 1858 von Graefe described the retinal changes which form an essential component of the retinitis pigmentosa group of disorders. By virtue of these observations his eponym is occasionally combined with that of Charles Usher of Aberdeen (1865–1942) *(see The Man Behind the Syndrome p. 177)* in the title of the condition comprising retinitis pigmentosa and progressive perceptive deafness.

Von Graefe's name is linked with that of Sjögren as a designation for an autosomal recessive syndrome of retinitis pigmentosa, perceptive deafness and spinocerebellar ataxia. There is some controversy concerning the syndromic boundaries of this condition, as Sjögren (1950) excluded retinitis pigmentosa, while Hallgren (1959) insisted upon the additional features of mental retardation and cataracts. For these reasons the condition is generally known by von Graefe's eponym with the addition of the names of either Hallgren or Sjögren.

# References

Hallgren B (1959) Retinitis Pigmentosa combined with congenital deafness; with vestibulo-cerebellar ataxia and mental abnormality in a proportion of cases. A clinical and genetico-statistical study. Acta Psychiat Scand 34(Suppl 138): 1–97

Leopold IH (1958) Albrecht von Graefe. Diabetes, NY 7 (2): 151–154

Sjögren T (1950) Hereditary congenital spinocerebellar ataxia accompanied by congenital cataract and oligophrenia. A genetic and clinical investigation. Confinia Neur, Basel 10: 293–308

Ullman EV (1954) Albrecht von Graefe: The man in his time. Am J Ophthalmol 38 Parts I–III: 525–809

Von Graefe A (1858) Exceptionelles Verhalten des Gesichtsfeldes bei Pigmententartung der Netzhaut. Graefe Arch Klin Exp Ophthal 4: 250–253

# VOORHOEVE, Nicolaas

*(1879–1927)*

*Courtesy*: D. Stewart, Librarian, Royal Society of Medicine, London

V OORHOEVE disease, or osteopathia striatum, is an autosomal dominant skeletal dysplasia in which parallel lines of sclerosis are radiographically evident in the pelvis and tubular bones.

# Biography

V OORHOEVE was a pioneer of Dutch radiology and an editor of the journal *Acta Radiologica.*

Nicolaas Voorhoeve was born in the Hague, Holland, on 15 January 1879. After receiving his schooling in that city, he qualified in medicine in 1905, at the University of Leiden. He was a diligent student with a critical scientific spirit, and at that early stage it was evident that he was destined for an academic career. Voorhoeve gained postgraduate experience in Berlin and Paris, before returning to Leiden in 1912 as assistant to Professor Pel. In this capacity Voorhoeve directed the radiology laboratory, obtaining further specialised experience with Wertheim Salomonson in Leiden and Albers-Schönberg in Hamburg (*see The Man Behind the Syndrome* p. 3). When Salomonson died in 1920, Voorhoeve was a natural candidate for the vacant chair; it was not until 1926, however, that he was finally appointed to the professorship. Voorhoeve's tenure was short-lived and he died in the following year.

Voorhoeve had a reputation as an excellent teacher of radiology, with vast knowledge and considerable oratorial skills. He was also a competent medical author and his main academic contribution consisted of a large number of articles which were published in the radiological literature. His skills in this direction served him well when he became editor of the prestigious journal *Acta Radiologica.*

Not a great deal has been written about Voorhoeve's character or personal life, but his friend, Dietz, with whom he had shared his student days, stated in his obituary "Voorhoeve was one of those men whom it was necessary to know for a long time to be able to appreciate fully their great qualities".

Voorhoeve developed carcinoma of the lung and died in Amsterdam on 8 August 1927.

# Nomenclature

I N 1924 Voorhoeve wrote an account of the radiological appearance of linear sclerotic striations in the pelvis and long bones of a father and his son and daughter. He believed that this was an original observation, commenting in the title of his article that these changes had not been previously described.

In the following year Sir Thomas Fairbank (*see The Man Behind the Syndrome* p. 55) reported a boy aged 12 years with similar but unilateral bony involvement and in 1935 he presented a case at a meeting of the Royal Society of Medicine, London, introducing the current descriptive title "osteopathia striatum". In his classical monograph *Atlas of General Affections of the Skeleton* Fairbank (1951) mentioned that he had discussed his patient with Voorhoeve, and that they had agreed that both reports pertained to the same entity.

In 1953 Hurt used Voorhoeve's eponym in conjunction with the term "osteopathia striatum" and postulated a relationship with osteopetrosis. Thereafter reports appeared regularly in the literature and it became evident that although osteopathia striata could occur in isolation, these radiological changes were often associated with other sclerosing bone dysplasias, and with conditions such as focal dermal hypoplasia. In 1978 Horan and Beighton investigated an affected girl in Cape Town, traced her family in the UK and documented autosomal dominant inheritance of osteopathia striata with cranial sclerosis. Further reports followed and the phenotype was widened; it now seems likely that a significant proportion of persons with supposedly isolated osteopathia striata actually have this disorder.

# References


Dietz PJ Ph (1927) Prof Nicolaas Voorhoeve 1879–1927. In Memoriam. Acta Radiol 8(4): 271–273

Fairbank HAT (1925) A case of unilateral affection of the skeleton of unknown origin. Br J Surg 12: 594–597

Fairbank HAT (1935) Osteopathia striata. Proc Roy Soc Med 28: 1616

Fairbank HAT (1951) *An Atlas of General Affections of the Skeleton.* Williams & Wilkins, Baltimore

Horan F, Beighton P (1978) Osteopathia striata with cranial sclerosis. An autosomal dominant entity. Clin Genet 13: 201–206

Hurt RL (1953) Osteopathia striata –Voorhoeve's disease. J Bone Joint Surg 35: 89–96

Voorhoeve N (1924) L'image radiologique non encore décrite d'une anomalie du squélette. Ses rapports avec la dyschondroplasie et l'osteopathia condensans disseminata. Acta Radiol, Stockholm 3: 407–427

# VROLIK, Willem
## *(1801–1863)*

*From*: Nederlands Tijdschrift voor Geneeskunde (1984) 128(32): 1531
*Courtesy*: Nederlands Tijdschrift voor Geneeskunde, Amsterdam

V ROLIK disease, or osteogenesis imperfecta congenita, is a potentially lethal disorder in which affected neonates present with multiple fractures due to skeletal fragility. The milder late-onset form of the condition, osteogenesis imperfecta tarda, sometimes bears the eponyms of Lobstein (p. 109) and van der Hoeve (p. 173).

# Biography

V ROLIK was a Dutch anatomist and physiologist who enjoyed a long and successful career at the Amsterdam Athenaeum during the nineteenth century.

Willem Vrolik was born on 29 April 1801 in Amsterdam and studied at the Amsterdam Athenaeum, where his father, Gerard Vrolik (1775–1859), was a professor. He subsequently continued his training at the University of Utrecht. Vrolik was academically gifted and received a prize for an essay written in Latin on the hearing mechanism in animals and man. He completed his studies in Paris and qualified in medicine in 1823 after defending his doctoral thesis on the influence of spinal malalignment on the circulation.

Vrolik commenced medical practice in Amsterdam, while continuing with his anatomical investigations and in 1828, at the age of 27 years, he obtained an academic post in anatomy and physiology at the University of Groningen. In 1830 Vrolik was commissioned into the Dutch army and took part in the campaign against Belgium. In the following year he accepted a call to a professorial chair at his alma mater, the Amsterdam Athenaeum, where he spent the next 32 years teaching anatomy, physiology, surgery and natural history. The Athenaeum contained a noted collection of anatomical specimens, especially skulls, which had been acquired by his father, and which bore his name. Vrolik documented the collection and added many specimens, thereby gaining an international reputation for the museum and for himself. In 1869 this valuable collection was purchased by the city of Amsterdam and it now reposes in the Department of Anatomy at Amsterdam University.

The fields of comparative anatomy and pathological anatomy embraced Vrolik's main research interests. In particular, he used specimens from the Vrolik Collection as a basis for an extensive review of mammalian and human teratology in his *Handbook of Pathological Anatomy* published in 1842, and in his *Tabulae ad Illustrandum*, which was printed in large format lithography in 1844–1849. He received international recognition for this work, including a prize from the French Academy of Science, and he is remembered for his contributions to the theory of teratology.

Vrolik was a devout Christian and a deacon of the Lutheran Church. His wife, Theodora Cornelia van Doorn, bore 2 sons and 5 daughters; sadly the latter all died in early childhood. Towards the end of his career Vrolik's health began to fail and he resigned from his chair in 1863, dying in Amsterdam in the same year.

# Nomenclature

O STEOGENESIS imperfecta (OI) is a heterogeneous disorder which was categorised by Sillence and colleagues in 1979 into 4 main types. Prior to this initiative OI had been divided in 1906 by Looser into the severe infantile, or congenita type, and the mild, tarda or late-onset type. The former was named after Vrolik and the latter bore the eponym of Lobstein and, sometimes, that of van der Hoeve.

In 1849 Vrolik published a treatise on embryogenesis and depicted the skeleton of an infant with multiple fractures in the long bones, and Wormian bones in the skull. This interesting drawing, which clearly represents the disorder which came to be known as "osteogenesis imperfecta congenita", has been reproduced in successive editions of *McKusick's Heritable Disorders of Connective Tissue*. In 1905 Porak and Durante described the same condition, using the misnomer "achondroplasia" and thereafter these authors' names were sometimes applied to the disorder, especially in the French literature. This format has now been abandoned.

The parents of infants with OI congenita are invariably unaffected and for many years it was thought that inheritance was probably autosomal recessive. This misconception has now been resolved by molecular techniques, which revealed that the majority of affected infants represent new dominant mutations. The elucidation of the molecular basis of OI has also permitted further subcategorisation and Vrolik's eponym, which remained in vogue for more than a century, is now rarely used.

(For further details of the nomenclature *see* Lobstein p. 109 and Van der Hoeve p. 173.)

# References


Baljet B (1984) Willem Vrolik als teratoloog. Ned Tijdschr Geneeskd 128(32): 1530–1534

Looser E (1906) Zur Kenntnis der osteogenesis imperfecta congenita et tarda. Mitt Grenzgeb Med Chir 15: 161–207

Porak C, Durante G (1905) Les micromélies congenitales. Achondroplasie vraie et dystrophie périostale. Nouv Icon Salpêtrière 18: 181–538

Sillence DO, Senn A, Danks DM (1979) Genetic heterogeneity in osteogenesis imperfecta. J Med Genet 16: 101–116

Vrolik W (1844–1849) *Tabulae ad illustrandam embryogenesin hominis et mammalium, tam naturalem quam abnormem.* Lipsiae, Weigel

# WALKER, Arthur Earl

*(1907–1995)*

D ANDY-WALKER syndrome, *see* Dandy, p. 49; Walker-Warburg syndrome, *see* Warburg, p. 187.

# Biography

W ALKER was the doyen of North American neurosurgeons. He occupied the prestigious chair of neurosurgery at the Johns Hopkins Hospital, Baltimore, USA and made many important academic contributions in his speciality during the middle period of the present century.

Arthur Earl Walker was born in Winnipeg, Manitoba, Canada in 1907 and qualified at the University of Alberta in 1930. After internship at Toronto Western Hospital he trained in neurology and neurosurgery at the University of Chicago, USA. At this stage in his career, Walker had already developed a profound interest in brain structure and function, and he was able to integrate the scientific aspects of neuropathology, electroencephalography and neurophysiology with clinical neurosurgery.

Walker was soon acknowledged as a leader of a new generation of scientifically orientated neurosurgeons; the award of a Rockefeller Fellowship facilitated further study at Yale, Amsterdam and Brussels, and in 1937 he became instructor in neurological surgery at the University of Chicago. When the USA entered World War II, Walker joined the army and was assigned to the neurology and neurosurgical service in charge of a post-traumatic epilepsy unit at Cushing General Hospital, Framingham, MA. In 1946 after the death of the great Walter Dandy, and at the completion of his military service, Walker was called to the chair of neurosurgery at the Johns Hopkins Hospital, Baltimore. He spent the most productive years of his career in this post, retiring in 1972 at the age of 65 years. His few spare moments were devoted to skiing, tennis and golf.

Walker made numerous contributions in many facets of neurosurgery; he elucidated the organisation of the spinothalamic tract, introduced new surgical procedures and wrote classical descriptions of congenital malformations, including the anomaly which bears his name. He then turned his attention to epilepsy and investigated the underlying electrophysiological processes. Walker was an innovator throughout his career and in his later years he introduced ultrasound, stereotactic surgery and histochemistry in the investigation and management of neurosurgical disorders. He published extensively, including more than 300 medical articles and several books. Walker's first monograph *The Primate Thalamus* appeared in 1938, when he was 31 years of age, and thereafter his output of publications continued undiminished throughout his career.

In addition to his investigative and literary skills, Walker was a noted teacher and his training programme attracted young surgeons from throughout the world. He was deeply involved in international neurosurgical affairs; Walker was a founder of the World Federation of Neurosurgical Societies in 1955 and he served as president from 1965 until 1969. He also held high office in several neurosurgical associations in North America and rendered notable service as historian and advisor in procedural matters.

Upon his retirement from the Johns Hopkins Hospital, Walker accepted a post as emeritus professor of neurology and surgery/neurosurgery at the University of New Mexico School of Medicine, Albuquerque, where he directed a national research programme on the criteria of brain death. In 1988 he published a 40-year follow-up of the patients with post-traumatic epilepsy whom he had seen at Cushing General Hospital in 1945–1946 and, in 1989, at the age of 82 years, he edited the English translation of a more than 700-page monograph on functional and stereotactic surgery by Professor E.I. Kandel of Moscow.

In his latter years, Walker's drive, intellect and academic output remained undiminished. He was regarded with respect tinged with awe, but nevertheless, he remained accessible to his peers and junior colleagues. Walker died suddenly from a heart attack on 1 January 1995 while travelling through Arizona on the way to California.

# Nomenclature

I N February 1942, Walker and John Taggart participated in a meeting of the Chicago Neurological Society and gave a presentation entitled *Congenital Atresia of the Foramens of Luschka and Magendie*. In this talk, which was published as a comprehensive review in the Archives of Neurology and Psychiatry, they mentioned 6 cases in the literature, including the child documented by Dandy and Blackfan in 1914 (*see The Man Behind the Syndrome* p. 19), and added 3 more cases from their own experience.

Benda (1954) employed the conjoined eponym "Dandy-Walker"; this format has stood the test of time and is still widely used to describe the anatomical abnormality of the foramina which leads to internal hydrocephalus. It is now recognised that this anomaly is non-specific and that it can be a component of several distinct syndromic entities (Chitayat et al. 1994).

In addition to the Dandy-Walker syndrome, Walker's name is used in conjunction with that of the Danish ophthalmologist, Mette Warburg (*see* Warburg, p. 187).

# References


Benda CE (1954) The Dandy-Walker syndrome or the so-called atresia of the foramen of Magendie. J Neuropath Exp Neurol 13: 14–39

Chitayat D, Moore L, Del Bigio MR, MacGregor D, Ben-Zeev B, Hodgkinson K, Deck J, Stothers T, Ritchie S, Toi A (1994) Familial Dandy-Walker malformation associated with macrocephaly, facial anomalies, developmental delay and brain stem dysgenesis: a new syndrome? Am J Med Genet 52: 406–415

Dandy WE (1921) Diagnosis and treatment of hydrocephalus due to occlusions of the foramina of Luschka and Magendie. Surg Gynecol Obstet 32: 112–124

Dandy WE, Blackfan K (1914) Internal hydrocephalus: an experimental, clinical and pathological study. Am J Dis Child 8: 406–482

Niedermeyer E (1995) In memoriam: A Earl Walker: neurosurgeon, neuroscientist, and epileptologist. Epilepsia 36(5): 516–521

Obituary (1995) Walker "Man of integrity". Albuquerque Journal, 8 January

Taggart J, Walker AE (1942) Congenital atresia of the foramens of Luschka and Magendie. Arch Neurol Psych 48: 583–612

Udvarhelyi GB (1968) Tribute to Dr A Earl Walker. J Nerv Ment Dis 147: 2–13

# WARBURG, Mette

*(b. 1926)*

W ALKER-WARBURG syndrome comprises liss-encephaly and occipital encephalocele, associated with cerebral midline defects, microphthalmia, congenital glaucoma and other variable ocular abnormalities. The condition, which is lethal in early infancy, is inherited as an autosomal recessive trait.

# Biography

W ARBURG is a Danish ophthalmologist with an international reputation for her contributions in the field of genetic eye disease.

Mette Warburg graduated from the University of Copenhagen, Denmark, in 1952, undertook postgraduate training in ophthalmology and obtained certification as a specialist in 1966. She held appointments as senior surgeon at the University Eye Clinic, Arrhus University and later at the eye clinics associated with the Royal Institute for the Blind, Copenhagen. In 1989 Warburg was head of the ophthalmological service for persons with multiple handicaps at the Gentofte Hospital medical school.

Warburg has lectured extensively in her own country and overseas and she has delivered the Doyne, Bjerrum and Franceschetti memorial lectures. Amongst her other achievements she has served as vice-chairman of the Danish Ophthalmological Society and the Medical Research Council of Denmark.

Warburg's research interests include congenital ocular anomalies, ophthalmogenetics and the aetiology of blindness in mentally retarded individuals. Her extensive publications include more than 140 articles on these topics. Warburg's doctoral thesis was devoted to an X-linked blindness syndrome, which she named "Norrie disease" after a Copenhagen ophthalmologist of an earlier era (*see The Man Behind the Syndrome* p. 125). Together with her co-workers she was subsequently able to identify the gene locus for this condition on the short arm of the X-chromosome.

Outside her professional life, Warburg devotes her time to her children and grandchildren, and for exercise, she enjoys gardening and hill walking.

# Nomenclature

I N 1942 Walker reported a girl aged 4 months with brain and eye abnormalities, hydrocephalus, agyria, microphthalmia and retinal dysplasia. He employed the term "lissencephaly" in the title of this article, in order to emphasize the abnormally smooth surface of the cerebral hemispheres. The same patient was subsequently mentioned by Krause (1946) under the designation "congenital encephalo-ophthalmic dysplasia" and by Reese and Straatsma (1958) in a paper on "retinal dysplasia".

Warburg (1971) reviewed the literature concerning hydrocephalus, microphthalmos and retinal non-attachment and suggested that the association of these abnormalities might constitute a new syndrome. In 1978 she documented an affected boy with consanguineous parents and proposed that the condition was an autosomal recessive disorder. Chemke et al. (1975) reported recurrence in a consanguineous family and Pagon et al. (1978) made a similar observation. The latter authors proposed the mnemonic "HARD ± E" syndrome (hydrocephalus, agyria, retinal dysplasia, with or without encephalocele). In 1981 Pagon and Clarren suggested that the mnemonic should be abandoned and that eponymic priority should be accorded to Warburg. Dobyns et al. (1985) subsequently drew attention to Walker's early,

detailed description and proposed the designation "Walker-Warburg syndrome".

Warburg (1987) pointed out that congenital muscular dystrophy was an important component of the disorder and concurred with the viewpoint that cerebro-oculo-muscular dystrophy and the Walker-Warburg syndrome were the same entity. The nomenclature and diagnostic criteria have been reviewed in detail by Dobyns et al. (1989); more than 60 cases have now been reported and the conjoined eponym has gained general acceptance.

# References


Chemke J, Czernobilsky B, Mundel G, Barishak YR (1975) A familial syndrome of central nervous system and ocular malformations. Clin Genet 7: 1–7

Dobyns WB, Kirkpatrick JB, Hittner HM, Roberts RM, Kretzer FL (1985) Syndromes with lissencephaly II: Walker-Warburg and cerebro-oculo-muscular syndromes and a new syndrome with type II lissencephaly. Am J Med Genet 22: 157–195

Dobyns WB, Pagon RA, Armstrong D, Curry CJR, Greenberg F, Grix A, Holmes LB, Laxova R, Michels VV, Robinow M, Zimmerman RL (1989) Diagnostic criteria for Walker-Warburg syndrome. Am J Med Genet 32: 195–210

Krause AC (1946) Congenital encephalo-ophthalmic dysplasia. Arch Ophthalmol 36: 387–414

Pagon RA, Clarren SK (1981) HARD ± E: Warburg's syndrome (letter to the editor). Arch Neurol 38: 66

Pagon RA, Chandler JW, Collie WR, Clarren SK, Moon J, Minkin SA, Hall JG (1978) Hydrocephalus, agyria, retinal dysplasia, encephalocele (HARD ± E) syndrome: an autosomal recessive condition. Birth Defects Orig Art Ser XIV(6B): 233–241

Reese AB, Straatsma BR (1958) Retinal dysplasia. Am J Ophthalmol 45(4, part 2): 199–211

Walker AE (1942) Lissencephaly. Arch Neurol Psychiat 48: 13–29

Warburg M (1971) The heterogeneity of microphthalmia in the mentally retarded. Birth Defects VII-3: 136–152

Warburg M (1978) Hydrocephaly, congenital retinal non-attachment and congenital falciform fold. Am J Ophthalmol 85: 88–94

Warburg M (1987) Ocular malformations and lissencephaly. Eur J Pediatr 146: 450–452

# WEILL, Georges
*(1866–1952)*

*From:* Survey of Ophthalmology (1974) 19(1): 31

*Courtesy:* The Williams & Wilkins Co., Baltimore

W EILL-MARCHESANI syndrome or the "spherophakia-brachymorphia syndrome" is an autosomal recessive disorder in which dislocation of the ocular lens is associated with stunted stature and brachydactyly.

# Biography

W EILL was a leading ophthalmologist at the University of Strasbourg, France, in the period between the two World Wars.

Georges Weill was born in Strasbourg, Alsace, in 1866 and graduated in medicine in that city during the period that the University was under German control. After his internships, Weill trained in ophthalmology under Stilling and became involved in research into colour vision. In 1919, after the armistice, a new French medical faculty was established. Weill became associate professor of ophthalmology and in 1928, at the age of 62 years, he succeeded to the chair.

Weill was a dextrous surgeon and travelled widely to learn new surgical techniques. He was a generous, warm-hearted man who created a pleasant and relaxed atmosphere in his clinic. His teaching sessions were very popular and he was affectionately known as "Papa" by his students. Weill was famous for his Grand Rounds which were held on Sunday mornings, thus interfering with his students' outdoor activities! He did not undertake a great deal of research or publish many articles but his booklet *Ophthalmology for the General Practitioner* gained favour.

Following his retirement in 1937 at the age of 71 years, Weill became interested in medical history. When World War II broke out he moved to Bordeaux, where his son was a university professor, and then to Clermont-Ferrand in the Free French Zone, where the University of Strasbourg had been transferred. After the war Weill returned to Strasbourg where he had a happy and healthy old age, continuing mountaineering until shortly before his death in 1952 at the age of 86 years.

# Nomenclature

I N 1932 Weill reported 8 patients with dislocation of the lens and other dysmorphic features. Several of these persons had arachnodactyly and were regarded as having the Marfan syndrome, but two were of stunted stature with short, stiff digits. Seven years later Marchesani of Munster described 2 additional families with ectopia lentis and stubby digits and this condition subsequently bore his name. Later, when Weill's earlier report was recognised, the conjoined eponym came into general use. An alternative descriptive designation "spherophakia-brachymorphia syndrome" was introduced by Kloepfer and Rosenthal in 1955 but the eponymic format is still preferred.

There has been some confusion between the names of Georges Weill and Adolph Weil, as the former has sometimes been spelt "Weil". Adolph Weil (1848–1816), a German physician, was associated with the elucidation of the pathogenesis of an acute form of leptospirosis which is conventionally termed "Weil disease or leptospirosis icterohemorrhagica". Jean Weill, who was associated with André Leri (1875–1930) in the delineation of dyschondrosteosis, is also a different person.

# References

Kloepfer HW, Rosenthal JW (1955) Possible genetic carriers in the spherophakia-brachymorphia syndrome. Am J Hum Genet 7: 398–425

Marchesani O (1939) Brachydaktylie und angeborene Kugellinse als Systemerkrankung. Klin Mbl Augenheilk 103: 392–406

Nordmann J (1975) Georges Weill, in Remembrances of Things Past. Survey of Ophthalmology 19(5): 295–298

Weill G (1932) Ectopie des cristallins et malformations générales. Ann Occul 169: 21–44

# WERNICKE, Carl

## *(1848–1904)*

*From*: Goldstein K. (1953) In: Webb Haymaker (ed) The Founders of
Neurology (1st edn)

*Courtesy*: Charles C. Thomas, Springfield, Illinois
Portrait from Mschr Psychiat 18:i (1905)

W ERNICKE disease, or encephalopathy, comprises mental confusion, paralysis of the external ocular muscles, nystagmus and ataxia. Although associated with alcoholism, thiamine deficiency in genetically predisposed persons is now regarded as the major determinant of the condition.

# Biography

W ERNICKE was a German neuropsychiatrist of the nineteenth century. He made notable contributions to knowledge concerning the pathological anatomy of the brain, especially in relation to aphasia.

Carl Wernicke was born in 1848 in Tarnowitz, Upper Silesia, and he qualified in medicine at the University of Breslau. He then obtained postgraduate experience with Neumann at Breslau and Westphal in Berlin, before spending six months with Meynert in Vienna. Shortly afterwards Wernicke published a small book on the cerebral location of aphasic syndromes which brought international recognition at an early age. In 1878 Wernicke entered private practice in Berlin, as a specialist in nervous diseases. He was able to pursue his academic interests in neuroanatomy and in 1883 he published a 3-volume text. Wernicke returned to Breslau in 1885 and in 1890 was elevated to the chair of psychiatry, which he occupied until shortly before his death in 1904.

Wernicke's main interest lay in establishing correlations between clinical abnormalities in the nervous system and the underlying histopathological changes in the brain. His early studies involved the convolutions of the cerebral cortex and he subsequently investigated the effects of disruption of the vascular supply to regions of the cerebellum and the medulla oblongata. In this period he documented the clinical manifestations of the form of haemorrhagic polyencephalitis, to which his eponym is now attached (vide infra). Wernicke's investigations led him to conclude that the psychoses were diseases of the brain and he regarded the specialities of neurology and psychiatry as being indivisible. Controversy on this issue continued throughout his life time.

A former pupil of Wernicke's described him as being "a taciturn and reserved man, not easy to deal with. He did not have much contact with his younger pupils but his way of examining patients and his demonstrations were so elucidating and stimulating that we who had the good fortune to attend his clinics were deeply influenced in our further consideration of neurological and psychiatric problems. We could never forget him. His influence can be seen in the work of each of his many pupils, not a small number of whom became men of stature in their own right in the profession."

Wernicke was called to the chair of psychiatry at the University of Halle in 1904. Shortly after his arrival he was killed in an accident, while riding a bicycle in the Thuringian forest.

# Nomenclature

I N the course of his investigations into lesions of the brain and their clinical consequences, Wernicke recognised a syndrome of confusion, oculomotor paralysis and ataxia, which was associated with petechial haemorrhages in the grey matter around the third and fourth ventricles and the aqueduct of Sylvius. In a description based upon three patients published in 1881, he termed this condition "acute superior haemorrhagic polioencephalitis". In articles which followed, the term "Wernicke encephalopathy" came into general use. The majority of affected persons were

alcoholics and the histopathological changes were thought to be the direct result of damage by alcohol abuse. Prior to Wernicke's publication, Gayet (1875) had described similar but more extensive intracerebral haemorrhage in a similar clinical setting. The conjoined eponym "Gayet-Wernicke syndrome" enjoyed brief favour but has now fallen into disuse.

In 1977 Blass and Gibson demonstrated an autosomal recessive deficiency of thiamine pyrophosphate binding factor, transketolase, in persons with Wernicke encephalopathy and in Korsakov psychosis. It was then recognised that alcohol-induced thiamine deficiency played a major role in these genetically susceptible individuals. In current concepts an alcoholic thiamine deficient person may have Wernicke encephalopathy in association with Korsakoff psychosis. In other words, Korsakoff pschysosis is the psychiatric manifestation of Wernicke disease. "Transketolase deficiency" and "alcohol-induced encephalopathy" are alternative designations (*see* Korsakov, p. 105).

In addition to the encephalopathy which bears his eponym, Wernicke's name is also associated with a specific type of aphasia and in the past it was also linked with a form of muscle cramp and a dementia. These conditions do not have a genetic basis.

# References

Blass JP, Gibson GE (1977) Abnormality of a thiamine-requiring enzyme in patients with Wernicke-Korsakoff syndrome. New Engl J Med 297: 1367–1370
Blass JP, Gibson GE (1979) Genetic factors in Wernicke-Korsakoff syndrome. Alcoholism Clin Exp Res 3: 126–134
Gayet M (1875) Affection encéphalique (Encéphalite diffuse probable). Localisée aux étages supérieurs des pédoncules cérébraux et aux couches optiques, ainsi qu'au plancher du quatrième ventricule et aux parois latérales du troisième. Observation recueillie. Arch Physiol Norm Path, Paris 2: 341–351
Obituary (Siemerling) (1905) Arch Psychiat, Berl 40: 1016–1019
Obituary (Ziehen) (1905) Mschr Psychiat 18: 1–4
Wernicke C (1881) Die acute hamorrhagische Polioencephalitis superior. In: *Lehrbuch der Gehirnkrankheiten.* Fischer, Kassel, pp. 229–242

# WHITE, Paul Dudley

## *(1886–1973)*

W OLFF-PARKINSON-WHITE syndrome is an electro-cardiographic abnormality in which a short PR interval is associated with a prolonged QRS complex. The syndrome may occur in isolation, as a familial trait, or with structural cardiac defects or cardiomyopathy. Supraventricular tachycardia is the usual presenting feature.

## Biography

W HITE was Professor of Medicine at the University of Harvard and the doyen of North American cardiologists.

Paul Dudley White was born in Roxbury, Massachusetts in 1886, where his father was a family doctor. He was educated at the Roxbury Latin Grammar School, proceeding to Harvard University and graduating in 1908. He obtained his medical degree at Harvard in 1913 and subsequently spent a year at University College Hospital medical school, London, financed by a Sheldon travelling scholarship. At the outbreak of World War I White volunteered for service with the British Forces and he spent the next two years in France. In 1917 he transferred to the American Expeditionary Force and after the armistice he joined the American Red Cross in Greece.

In 1919 White joined the staff of the Massachusetts General Hospital, Boston and embarked upon a career in cardiology. He became Professor of Medicine and with his outstanding clinical abilities established an international reputation for himself and his department. His classical monograph *Heart Diseases*, first published in 1931, went into several editions and set the seal on his career. In 1948 he was elected as president of the International Society of Cardiology and was subsequently president of the first World Congress of Cardiology. White retained his close contact with clinical cardiology and in the following years he attended President Eisenhower when he suffered a heart attack during his term of office as President of the United States of America.

White was an indefatigable traveller and he devoted considerable effort to fostering the speciality of cardiology and providing impetus for the establishment of units and institutes throughout the world. He was still active in old age, paying an academic visit to China in his 83rd year. White was a great believer in the importance of physical exercise; his slim physique was testimony to the fact that he kept to his principles of "walk more, eat less, sleep more".

White died in 1973, replete with honours, at the age of 85 years.

## Nomenclature

I N 1930, together with Wolff[1] and Parkinson, White co-authored a paper entitled *Bundle-branch block with a short P-R interval in healthy young people prone to paroxysmal tachycardia* (*see* Parkinson, p. 127). This disorder turned out to be comparatively common and numerous reports followed; the triple eponym enjoyed immediate favour and it has continued to be widely accepted.

With the accumulation of case reports, it has become evident that the Wolff-Parkinson-White (WPW) syndrome is very heterogeneous. It can exist with or without concomitants such as structural cardiac malformations and cardiomyopathy and the isolated form is often familial. In 1959 Harnischfeger drew attention to hereditary occurrence, and in 1978 Gillette and colleagues proposed that the condition had an autosomal dominant mode of inheritance. The concept of autosomal dominant inheritance of pre-excitation and accessory atrioventricular pathways in the context of the

WPW was introduced in 1987 by Vidaillet and colleagues. The harmonious triple eponym, which trips easily off the tongue, is firmly established and, unlike many other disorders which bear the name or names of their originators, there is no trend towards substitution of a descriptive title.

## References

Gillette PC, Freed D, McNamara DG (1978) A proposed autosomal dominant method of inheritance of the Wolff-Parkinson-White syndrome and supraventricular tachycardia. J Pediat 93: 257–258

Harnischfeger WW (1959) Hereditary occurrence of the pre-excitation (Wolff-Parkinson-White) syndrome with re-entry mechanisms and concealed conduction. Circulation 19: 28–40

Obituary (1973) Br Med J 4: 362.

Schneider RG (1969) Familial occurrence of Wolff-Parkinson-White syndrome. Am Heart J 78: 34–36

Vidaillet HJ Jr, Pressley JC, Henke E, Harrell FE Jr, German LD (1987) Familial occurrence of accessory atrioventricular pathways. New Engl J Med 317: 65–69

Wolff L, Parkinson J, White PD (1930) Bundle-branch block with short P-R interval in healthy young people prone to paroxysmal tachycardia. Am Heart J 5: 685–704

---

[1]Louis Wolff was born in Boston in 1898. He qualified in medicine at Harvard in 1922 and undertook his internship at the Massachusetts General Hospital. He specialised in cardiology and became visiting physician and chief of the electrocardiographic laboratory at the Beth Israel Hospital. Wolff died on 28 June 1972, at the age of 74 years. (Obituary (1972) Wolff, Louis. JAMA 220: 1755.)

# WILLI, Heinrich

*(1900–1971)*

*Courtesy*: Professor D. Klein, Switzerland

P RADER-LABHART-WILLI syndrome consists of obesity, low intelligence, small hands and feet, a characteristic facies and hypogonadism. A significant proportion of affected persons have a deletion in the long arm of chromosome 15.

# Biography

W ILLI was a distinguished Swiss paediatrician who had a productive career in Zurich during the twentieth century.

Heinrich Willi was born on 4 March 1900 in Chur, where he was the sixth of 9 children. He studied medicine at the University of Zurich, qualifying in 1925, and became a resident at the Zurich Institute of Pathological Anatomy and in the department of medicine at the Winterthur Hospital. In 1928 Willi moved to the Children's Hospital, Zurich, where he trained in paediatrics under the great Guido Fanconi (1892–1979), becoming assistant medical director in 1930. Willi obtained his doctorate in 1936 with a thesis on "childhood leukaemias" and in 1937 he was elected as successor to Professor Bernheim-Karrer, the first practicing specialist paediatrician in Zurich, and as director of the cantonal "Säuglingsheim Rosenberg", which later became the University Hospital for Neonatology. His main duties involved the care of neonates and premature infants and he made important contributions to the development of the speciality of neonatology. Willi remained in this post until his retirement in 1970, having occupied this position for more than 30 years.

Willi was always fascinated by medical research and he undertook investigations throughout his career, despite a very heavy clinical workload. His studies included dietary experiments, which he conducted on himself, the haematological effects of ascariasis and the delineation of the leukaemias of childhood; his work in this latter field achieved international recognition. Willi was an innovator and, as such, he was one of the first to recognise the importance of bone marrow aspiration as an investigative technique. In the later stages of his career Willi concerned himself with neonatal pathology, including the delineation of the condition which bears his name. At this time he also contributed to the understanding of developmental abnormalities in infants born to diabetic mothers. Willi's achievements led to his election as president of the Swiss Association for Paediatrics 1959–1962 and he also received the honour of a nomination for the exclusive German Scientific Society, Leopoldina.

Apart from his research activities and routine duties, Willi managed a large private practice in which he was highly regarded as a conscientious and knowledgeable paediatrician by his colleagues and his patients. He was still working in his practice in Zurich when he died suddenly on February 16th 1971, at the age of 71 years.

In his obituary, Willi's former mentor, Professor Guido Fanconi paid tribute to his qualities. "He was a kind and blessed helper to his little patients and their fearful parents. He is mourned not only by the many pupils and his elderly teachers but also by innumerable parents and previous patients, now also adult. All mourn the dear, intelligent person who understood how to be researcher, teacher and a good doctor at the same time."

# Nomenclature

I N 1956, 3 Swiss investigators, Prader, Labhart and Willi described 9 children with the tetrad of small stature, mental retardation, obesity and small hands and feet. Prader and Willi reviewed the condition in 1961, expanded the phenotype and drew attention to the presence of hypotonia in infancy and the development of diabetes mellitus in later childhood (*see* Labhart, p. 214; *The Man Behind the Syndrome* Prader, p. 223).

An early account of the Prader-Labhart-Willi syndrome can be found in a monograph on mongolism, in which Brain (1967) cites a case description made by Down (1887) in his classical text *Mental Affections of Childhood and Youth*. Using the term "polysarcia" Down documented details of an obese mentally retarded girl aged 14 years, who was 4 feet 4 inches in height and weighted 196 pounds. At the age of 25 years she weighed 210 pounds. Down stated "Her feet and hands remained small and contrasted remarkably with the appendages they terminated. She had no hair in the axillae and scarcely any on the pubis. She never menstruated nor did she exhibit the slightest sexual instinct".

Once the disorder had been delineated, numerous reports followed and it became evident that the syndrome was comparatively common. Indeed, by 1986 the Prader-Willi Association of the USA and Canada had a register of 1,595 affected persons. An interstitial deletion in the long arm of chromosome 15 has been recognised in a significant proportion of affected persons and a relationship with the Angelman syndrome, mediated by the mechanism of "imprinting", has been proposed (*see* p. 13).

Following the original paper of Prader, Labhart and Willi the eponymic title was usually employed in the reports of the condition. There is now a tendency to drop the name "Labhart" and to designate the disorder "Prader-Willi syndrome (PWS)". Nevertheless, some authors, moved by generosity of spirit or an awareness of historical precedents, have chosen to retain the triple eponym, Prader-Labhart-Willi (PLWS).

# References


Brain RT (1967) Historical introduction. In: Wolstenholme GEW, Porter R (eds). *Mongolism*. Little, Brown and Co., Boston, pp. 1–5

Down JL (1887) *Mental Affections of Childhood and Youth*. Churchill, London, p. 172

Fanconi G (1971) Prof Heinrich Willi zum Gedenken. Neue Bündner Zeitung, 27 März: 3

Prader A, Labhart A, Willi H (1956) Ein Syndrom von Adipositas, Kleinwuchs, Kryptoorchismus und Oligophrenie nach Myatonieartigem Zustand im Neugeborenenalter. Schweiz Med Wschr 86: 1260–1261

Prader A, Willi H (1961) Das Syndrom von Imbezillität, Adipositas, Muskelhypotonie, Hypogenitalismus, Hypogonadismus und Diabetes mellitus mit "Anamnese. 2nd Int Congr Ment Retard, Vienna. Karger, Basel & New York, Part 1: 353

Wissler H (1971) Zum Hinschied von Prof. Dr. med. Heinrich Willi. Rade zur abdankungs – feier am 24 Februar

Zellweger H, Schneider HJ (1968) Syndrome of hypotonia-hypomentia-hypogonadism-obesity (HHHO) or Prader-Willi syndrome. Am J Dis Child 115: 588–598

# WOLCOTT, Carol Nancy Dettman

*(1941–1994)*

W OLCOTT-RALLISON syndrome comprises early onset diabetes mellitus and stunted stature due to multiple epiphyseal dysplasia. Inheritance is autosomal recessive.

# Biography

W OLCOTT was a North American pediatrician and a proponent of women's health care.

Carol Nancy Dettman Wolcott was born near Chicago in 1941 and spent her childhood in Minneapolis, Minnesota. She had considerable musical talents and contemplated a career in this field but, influenced by family members, she studied medicine at the University of Minnesota. After qualification, Wolcott trained in paediatrics at the University of Utah and while working with Professor Rallison in the care and management of juvenile diabetes, she was involved in the delineation of the syndrome that bears her name.

Wolcott moved to the University of Wisconsin where she made significant innovations in the teaching programme for paramedical colleagues. Her next move was to the University of Nebraska where she changed emphasis and became involved with women's health care issues. Through this initiative she was able to bring awareness of sexually transmitted diseases to the adolescent population.

Her musical talents continued to develop and she gave vocal performances in Nebraska and Connecticut. Wolcott had planned to obtain a postgraduate degree in performing arts but her ambitions were frustrated by her untimely death in 1994. She is survived by her husband George J. Wolcott, a paediatric neurologist of Lincoln, Nebraska.

# Nomenclature

I N 1972, in conjunction with her departmental head, Professor ML Rallison, Wolcott documented 2 brothers and a sister with insulin dependent diabetes mellitus, of infantile onset, plus multiple epiphyseal dysplasia. In their article, Wolcott and Rallison stated: "We have recently studied a family with three children who had onset of diabetes mellitus in early infancy; multiple epiphyseal dysplasia and abnormalities of the teeth and skin were also observed. The association of multiple epiphyseal dysplasia with infancy-onset diabetes mellitus has never been reported to our knowledge and suggested to us that we might be dealing with a new familial syndrome."

They went on to give the following case description: "Patient JSS, a boy, was hospitalized at 8 weeks of age with the history of frequent urination of 2 weeks' duration. The infant's urine showed 4+ glucosuria. The blood sugar concentration was 500 mg per 100 ml. The diabetes was stabilized with 10 units of Lente insulin daily. Later hospitalizations were required for treatment of hypoglycemia, diabetic ketoacidosis, and three fractures of the femurs, which occurred over a 6 month period (two on the right and one on the left). Radiographs taken at the time of the child's first fracture showed diffuse emineralization of the bones.

The child was admitted to the University of Utah Medical Center at 3 years of age because of the multiple fractures. An odd, stiff-kneed gait was noted, along with muscular atrophy and weakness of both legs, thought to be secondary to disuse and prolonged immobilization for treatment of the fractures. There was full range of motion in both hips, but limitation of knee flexion and elbow extension bilaterally. In addition, a brown mottling of the teeth near the gums and a cardiac murmur, thought to be functional, were noted. He was receiving 11 units of NPH insulin daily with supplements of regular insulin. Radiographs of the pelvis and extremities showed diffuse demineralization. The epiphyses were small, irregular, and poorly calcified. There was prominent beaking of the metaphyses of the distal femora and proximal radii. There was also a tubulation defect of the metacarpals and phalanges. Radiographs of the skull and chest were normal.

Biopsy of the iliac crest showed the cartilage cells at the epiphyseal line to have a narrow zone of vacuolation. The zone of provisional calcification was practically nonexistent. Bone trabeculae of irregular width and orientation abutted directly onto the hypoplastic epiphyseal line. The bone trabeculae in the metaphysis contained small foci of incomplete ossification. The biopsy and radiologic findings were consistent with the diagnosis of multiple epiphyseal dysplasia."

The eponym"Wolcott-Rallison" was employed by Stoss et al. (1982) in an article in which they drew attention to the involvement of the spine. The condition is rare and by 1995 only 7 cases had been reported.

# References

Stoss H, Pesch, H-J, Pontz B, Otten A, Spranger J (1982) Wolcott-Rallison syndrome: diabetes mellitus and spondyloepiphyseal dysplasia. Europ J Pediat 138: 120–129

Wolcott CD, Rallison ML (1972) Infancy-onset diabetes mellitus and multiple epiphyseal dysplasia. J Pediat 80: 292–297

# WOLMAN, Moshe

*(b. 1914)*

Woman disease, also known as primary familial xanthomatosis or lysosomal acid lipase deficiency, is a progressive, lethal disorder of infancy characterised by accumulation of lipid material in the viscera.

# Biography

WOLMAN had a long and productive career as an academic pathologist in Israel in the decades following World War II.

Moshe Wolman was born in Warsaw on 19 October 1914. His family immigrated to Palestine (present day Israel) in 1925 and, after completing his schooling in that country, he studied medicine in Florence and Rome. In 1938 he became a pathologist at the Hebrew University and Hadassah Hospital, Jerusalem. Following the outbreak of World War II Wolman volunteered for the British Army and served in the Middle East and East Africa, initially in the 101 Mission (Gideon's Force) in Ethiopia, under the command of Wingate. In 1946, after the armistice, he rejoined the department of pathology at the Hadassah Hospital and in 1949 received an additional appointment at the new medical school in Jerusalem. Wolman moved to Tel Aviv in 1959 and became head of pathology at the Tel Hashomer Hospital, where he founded the department of cell biology and pathology. In 1964, at the inception of the Tel Aviv University Medical School, he received professorial status. Wolman retired from the hospital in 1979 and from the University in 1985, becoming professor emeritus.

Wolman is author or editor of 5 books, including a text written in Hebrew on the subject of demagoguery and rhetoric. He has also published more than 200 articles in the medical literature and his scientific contributions, made during 5 decades, have been recognised by several distinctions, including membership of the Leopoldina Academy and the award of the Pearse Prize in Histochemistry from the Royal Microscopical Society.

In retirement, Wolman lived in Tel Aviv. He maintained his academic connections, continuing his investigations and contributions; in 1992, at the age of 80 years he promulgated his observation on the positive response to a form of percutaneous therapy in Wolman Disease.

# Nomenclature

IN correspondence with the authors, Wolman gave the following account of the delineation of the condition which now bears his name: "In anecdotal fashion the way in which the disease was discovered is as follows. My colleague and friend Dr Schorr, a radiologist, called my attention to the strange case of a deceased infant with calcified adrenals. At that time (also before and since), I was interested and active in lipid histochemistry. At autopsy the findings were striking and I was faced with the fact that a similar case had been reported in the standard US radiology book as Niemann-Pick disease (NP). Another case was published by a New Zealander also as NP disease. The last mentioned author based his diagnosis on a false assumption: the stored material was frankly sudanophilic. I was convinced on the basis of my experience that frank sudanophilia which I observed in the autopsied infant is inconsistent with storage of phospholipids.

We wrote to the American radiologist asking him how he knew his case was one of NP disease. He answered that that was the answer he got from his pathologist. I then performed various histochemical procedures on the stored material and found that not all storing cells contained phospholipids. I now felt sure that this was not a case of NP, as cells affected by the enzyme deficiency which accumulated metabolites must store the compound which cannot be metabolized, with or without accompanying materials. As the material stored in many foam cells did not contain phospholipids the primary defect could not be sphingomyelin.

The problem then was to fight prevailing opinion among our own clinicians, and world authorities. I was helped then (just at the opening of a clinico-pathological discussion on the case) by the findings of the biochemist who confirmed my histochemical findings. Materials from autopsies of two additional sibs were forwarded to me after I moved from Jerusalem. The findings were similar and were again reported.

I think that my name was given to the condition by Dr Crocker and his associates in Boston. I believe that the association is justified, as I proposed that this was not NP disease and for a period was alone in upholding this notion."

In 1966, the year after Crocker and colleagues had established the eponym, the term "Wolman disease" was used in the Japanese literature by Konno et al. and by Speigel-Adolf et al. in a report of 3 affected siblings in the USA. Since that time, many cases have been documented and Wolman's eponym is firmly established.

A condition termed "Wolman disease with hypolipoproteinemia and acanthocytosis" was described by Eto and Kitagawa in 1970. The genetic basis of classical Wolman disease has been elucidated (Anderson et al. 1993), and it is anticipated that the separate syndromic identity or otherwise of this variant form will be resolved by molecular studies.

# References


Anderson RA, Rao N, Byrum RS, Rothschild CB, Bowden DW, Hayworth R, Pettenati M (1993) In situ localization of the genetic locus encoding lysosomal acid lipase/cholesterol esterase (LIPA) deficient in Wolman disease to chromosome 10q23.2-q23.3. Genomics 15: 245-247

Crocker AC, Vawter GF, Neuhauser EBD, Rosowsky A (1965) Wolman's disease: three new patients with a recently described lipidosis. Pediatrics 35: 627-640

Eto Y, Kitagawa T (1970) Wolman's disease with hypolipoproteinemia and acanthocytosis: clinical and biochemical observations. J Pediatr 77: 862-867

Konno T, Fujii M, Watanuki T, Koizumi K (1966) Wolman's disease: the first case in Japan. Tohoku J Exp Med 90: 375-389

Spiegel-Adolf M, Baird HW, McCafferty M (1966) Hematologic studies in Niemann-Pick and Wolman's disease (cytology and electrophoresis). Confin Neurol 28: 399-406

Wolman M, Sterk VV, Gatt S, Frenkel M (1961) Primary family xanthomatosis with involvement and calcification of the adrenals: report of two more cases in siblings of a previously described infant. Pediatrics 28: 742-757

# WORTH, Harry Mullins

*(1897–1987)*

**W**ORTH endosteal hyperostosis is an autosomal dominant sclerosing bone dysplasia characterised by moderate overgrowth of the mandible with variable paralysis of the seventh and eighth cranial nerves. Radiological density of the skeleton is increased maximally in the calvarium.

# Biography

**W**ORTH graduated in dentistry and medicine in England and specialised in radiology. He spent the latter part of his career in Canada, where he delineated the syndrome which bears his name.

Harry Mullins Worth was born in England in June 1897. He graduated in dentistry in 1919 and in medicine in 1925, obtaining a diploma in medical radiology and electrology from the University of Cambridge in 1926. Worth was then appointed as a radiologist at Guy's Hospital, London, his alma mater, where he remained until 1939. During this period he was honorary radiologist to the West London Hospital, the Princess Louise Hospital for Children, Southwark Hospital and Epping Cottage Hospital and he also served as consulting radiologist to the Royal Air Force.

Worth moved to North America in 1939 and gained further experience at various centres, including the Peter Bent Brigham Hospital in Boston, the New York Memorial Hospital for Cancer and the Mayo Clinic. In 1942 Worth accepted an invitation to set up a postgraduate course in radiology in Toronto, Canada, for military medical officers and he was subsequently appointed as radiologist at the Wellesley Hospital in that city. He also lectured at the Toronto Dental College and in 1975 he received the honorary degree of LLD from the University of Toronto. In 1951 Worth moved to British Columbia, where he continued to lecture at the Dental School of the University of Washington, Seattle. In 1985 he finally retired at the age of 87 years.

Worth met his future wife at Guy's Hospital, London, during her training as a physiotherapist. She had been brought up in Durban, South Africa, where her father was a dentist; Worth also taught her 3 brothers, who were medical students at Guy's Hospital. His only child, Ann, qualified in medicine and became a pathologist at the Cancer Institute, Vancouver.

In 1987, at the age of 90 years, Worth was living in retirement in Vancouver, Canada. He retained an active interest in his speciality and, at the time of his death in 1987, he was still receiving radiographs from many quarters for his expert opinion.

# Nomenclature

**I**N the post-war era, when Worth was serving as radiologist at the Toronto General Hospital, he maintained personal and professional links with D. G. Wollin[1], whom he had taught during the war years. Wollin had moved to Kingston, Ontario, where he encountered patients with the sclerosing bone dysplasia which was subsequently known by Worth's name. The affected individuals were of early Canadian stock and after collecting radiographic material, Wollin sought Worth's collaboration. Their findings were published in 1966 under the title *Hyperostosis corticalis generalisata congenita*. There was some semantic confusion at this time, as in 1955 van Buchem and colleagues in Holland had published an account of a similar but more severe disorder, which they had termed "hyperostosis corticalis generalisata familiaris" (*see The Man Behind the Syndrome* p. 179). This condition, which had its origin in an endogamous Dutch community on the island of Urk in the Zuider Zee, could be differentiated from the disorder documented by Worth and Wollin by virtue of the respective autosomal recessive and autosomal dominant modes of inheritance.

The nosological difficulties were partially resolved in 1971 when Maroteaux and colleagues introduced the term "autosomal dominant osteosclerosis" for the disorder delineated by Worth and Wollin. Other authors, including Gelman (1977), Gorlin and Glass (1977) and Perez-Vicente and colleagues (1987) used the title in further reports of affected families. In 1974 Spranger and colleagues, in their classical *Atlas of Bone Dysplasias*, employed the designation "endosteal hyperostosis" to embrace both disorders while emphasising their separate syndromic identity by the addition of the eponyms "van Buchem" and "Worth". This format was subsequently codified in the May 1982 revision of the International Nomenclature of Constitutional Diseases of Bone (Paris Nomenclature) and it has been retained in the most recent version, the 1992 Bonn Nosology. Despite his contribution, the name of Worth's co-author, Wollin, is now rarely, if ever, used for the disorder.

# References


Gelman MI (1977) Autosomal dominant osteosclerosis. Radiology 125: 289–296

Gorlin RJ, Glass L (1977) Autosomal dominant osteosclerosis. Radiology 125: 547–548

Maroteaux P, Fontaine G, Schorfman W, Farriaux J-P (1971) L'hyperostose corticale generalisée a transmission dominante. Arch Franc Pediat 28: 685–698

Perez-Vicente JA, Rodriguez De Catro E, Lafuente J, Mateo MMD, Gimenez-Roldan S (1987) Autosomal dominant endosteal hyperostosis: report of a Spanish family with neurological involvement. Clin Genet 31: 161–169

Spranger JW, Langer LO, Wiedemann HR (1974) Bone dysplasias. In: *An Atlas of Constitutional Disorders of Skeletal Development*. Saunders, Philadelphia, p. 331

Worth HM, Wollin DG (1966) Hyperostosis corticalis generalisata congenita. J. Canad Assoc Radiol 17: 67–74


---

[1]Dr D.G. Wollin was born in Ingersoll, a suburb of London, Ontario, Canada, and trained in medicine at the University of Western Ontario. During World War II he served in the Royal Canadian Army Medical Corps and in 1943 he met his future collaborator, Worth, while attending a course in diagnostic radiology at the Toronto General Hospital. He remained in the speciality of radiology, moving at the end of the war to Montreal and then, in 1948, to Kingston, Ontario. He became director of diagnostic radiology at the Kingston General Hospital, Ontario, and enjoyed the status of associate professor in the faculty of medicine, Queens University.

# Section II

# Brief Biographies

# Aase, J.M.

Aase syndrome comprises triphalangeal thumbs and congenital haemolytic anaemia. Inheritance is probably autosomal recessive.

Jon Aase was born in Wisconsin and raised in California and Alabama. He graduated from Pomona College in Claremont, California, and received his medical qualification from Yale University in New Haven, Connecticut. After a paediatric internship at the University of Minnesota and residency at the University of Washington, Aase spent 2 years studying the indigenous population of Alaska as a fellow with the National Institute of Child Health and Human Development. He then returned to Seattle for a 2-year fellowship in dysmorphology with the late Dr David W. Smith. After 5 more years in Alaska, practising dysmorphology and general paediatrics, Aase joined the faculty of the University of New Mexico, where he served as chief of the dysmorphology division for 16 years. His bibliography contains more than 45 articles and a textbook, *Diagnostic Dysmorphology*, published in 1990.

Aase is currently clinical professor of paediatrics at the University of New Mexico and a consulting dysmorphologist in private practice. He also provides clinical and educational services for various State and national agencies. Aase lives in Albuquerque with his wife Kathleen and their 3 children.

## Reference

Aase JM, Smith DW (1969) Congenital anemia and triphalangeal thumbs: a new syndrome. J Pediatr 74: 471–474

# Aberfeld, D.C.

Schwartz-Jampel-Aberfeld syndrome comprises myotonia, blepharophimosis, stunted stature and joint contractures. Inheritance is autosomal recessive

Donald Aberfeld was born on 5 September 1932, in Bucharest, Romania. He qualified in medicine in 1957 at the University of Bucharest and spent a year as a physician at the neuropsychiatric clinic of the University of Vienna.

Following a neurological residency and fellowship in New York, Aberfeld obtained appointments as visiting neurologist at several hospitals in that city. In 1965, while associated with the Maimonides Hospital, Aberfeld was the senior author of a report concerning a brother and sister with the condition which now bears his name. He is at present an attending neurologist on the staff of St Vincent's Medical Center in New York City.

(See *The Man Behind the Syndrome* Jampel, p. 213.)

## Reference

Aberfeld DC, Hinterbuchner LP, Schneider M (1965) Myotonia, dwarfism, diffuse bone disease and unusual ocular and facial abnormalities (a new syndrome). Brain 88: 313–322

# Aicardi, J.

Aicardi syndrome consists of mental retardation, infantile spasms and chorioretinal abnormalities together with structural abnormalities of the brain and specific electroencephalographic changes. Inheritance is probably X-linked dominant, with male lethality.

Jean Aicardi was born in 1926 in Rambouillet, France. He qualified in medicine at the University of Paris and spent his whole career as a paediatric neurologist, with a special interest in paediatric epilepsy, at the Saint-Vincent de Paul and Enfants-Malades Hospitals in Paris. In 1955 Aicardi received a doctorate for his thesis entitled *Convulsions in the First Year of Life*. Thereafter his interest in this field led to the recognition of 2 girls with the disorder which now bears his name. In the years that followed he was able to accumulate more cases, all of whom were females. His senior colleague, Professor Jacques Lefebre, a co-author of an earlier abstract, encouraged him to delineate the new syndrome and in 1969 a full account was published in the French literature.

After his retirement in 1991, Aicardi took up appointments as honorary professor of child neurology at the Institute of Child Health and honorary consultant at the Hospital for Sick Children, London.

Aicardi married Jeanne Couturier in 1958.

## Reference

Aicardi J, Chevrie JJ, Rousselle F (1969) Le syndrome spasmes en fléxion, agénésie calleuse, anomalies chorio-rétiniennes. Arch Fr Pédiatr 26: 1103–1120

# Alagille, D.

Alagille syndrome, or arteriohepatic dysplasia, comprises intrahepatic cholestasis, a characteristic facies and variable visceral and skeletal malformations.

Daniel Alagille was born in Paris on 24 January 1925. After qualifying in medicine at the University of Paris he obtained a diploma in biochemistry and trained in paediatrics. He was appointed as associate professor of paediatrics in 1963 and elevated to full professorial status at the South Paris University in 1971. In this latter capacity Alagille founded a paediatric liver unit at the Bicêtre Hospital, where he served until his retirement in 1990.

Alagille has published more than 500 articles, together with several textbooks, including a classical monograph on liver disease in the infant, which has been translated into several languages. He was chief editor of the *Revue Internationale d'Hépatologie* (1954–1971) and of *Archives Françaises de Pédiatrie* (1964–1990).

Alagille has travelled extensively and lectured in many countries around the world. In his spare time he pursues his hobby of aviation, having held a pilot's licence for more than two decades. Alagille's contributions have been recognised by the bestowal of several honours and awards, including Chevalier de l'Ordre National du Mérite (1967) and Chevalier de la Légion d'Honneur (1988).

## Reference

Alagille D, Odièvre M, Gautier M, Dommergues JP (1975) Hepatic ductular hypoplasia associated with characteristic facies, vertebral malformations, retarded physical, mental and sexual development, and cardiac murmur. J Pediatr 86: 63–71

# Bannayan, G.A.

Bannayan syndrome is an autosomal dominant disorder comprising macrocephaly, growth retardation, lipomatosis and angiomatosis.

George A Bannayan was born on 15 December 1931 in Jerusalem, of Armenian ancestry. He obtained a bachelors degree in 1953 and a medical qualification in 1957 from the American University of Beirut. He then undertook postgraduate training in pathology at the Johns Hopkins Hospital, Baltimore, USA and the Memorial Hospital for Cancer and Allied Diseases. Thereafter, Bannayan became professor of pathology and head of anatomic pathology at the University of Texas Health Science Center at San Antonio where he remained until 1982. In 1994 he was residing in San Antonio, Texas, with his wife Odette, where he was the medical director of laboratories at San Antonio Regional Hospital, and the Women's and Children's Hospital. He also maintained a clinical pathology professorship at the University of Texas Health Science Center at San Antonio.

Bannayan's main academic interests are in renal, gastrointestinal and transplant pathology. He has authored and coauthored 31 publications and 12 chapters in the field of kidney disease.

## Reference

Bannayan GA (1971) Lipomatosis, angiomatosis and macrencephalia: a previously undescribed congenital syndrome. Arch Path 92: 1–5

# Beals, R.K.

Beals syndrome, or congenital contractural arachnodactyly, comprises a marfanoid habitus, digital contractures and crumpled external ears. Inheritance is autosomal dominant.

Rodney K Beals graduated from the University of Oregon Medical School in 1956. He trained as an orthopaedic surgeon and in 1961 became a member of the faculty at the Oregon Health Sciences University. Since 1981 he has been head of the division of orthopaedics and rehabilitation.

Beals has a special academic interest in the field of genetic connective tissue disorders, and in the early 1960s he developed a "Growth Clinic" where patients with skeletal dysplasias and other growth disturbances are seen. His work in this direction led to his selection as a travelling fellow of the American, British and Canadian Orthopaedic Associations in 1970. Beals is a regular contributor to the literature of heritable osseous dysplasias and syndromes, as well as general orthopaedic topics.

## Reference

Beals RK, Hecht F (1971) Congenital contractural arachnodactyly. J Bone Jt Surg 53: 987–993

# Beemer, F.A.

Beemer short rib syndrome is a rare dwarfing skeletal dysplasia, which is lethal in the neonatal period. Inheritance is autosomal recessive.

Frederikus Antonius Beemer was born on 18 October 1944 in Zwolle, Holland. He obtained a degree in biochemistry at the University of Utrecht and then studied medicine, qualifying in 1971. After a period of military service, he trained in paediatrics at the Binnengasthuis, Amsterdam, registering as a specialist in 1976. He was then appointed to the hospital staff, with responsibilities in the field of the inborn errors of metabolism. In 1977 Beemer took up his current post at the Utrecht University Children's Hospital, Het Wilhelmina Kinderziekenhuis, where he was the co-founder of the Clinical Genetics Centre for Utrecht. Beemer became head of professional education and medical director of this Centre in 1990. He is the author of more than 150 publications, mainly in the fields of dysmorphology, skeletal dysplasias and inborn errors of metabolism.

Professor Beemer is married and has 6 children.

## Reference

Beemer FA, Langer LO Jr, Klep-de Pater JM, Hemmes AM, Bylsma JB, Pauli RM, Myers TL, Haws CC III (1983) A new short rib syndrome: report of two cases. Am J Med Genet 14: 115–123

# Bernard, J.

Bernard-Soulier disease is a congenital bleeding disorder characterised by giant platelets and thrombocytopenia.

Jean Bernard is a distinguished French haematologist and medical scientist and for many years he was professor of haematology and director of the Institute for Leukemia at the University of Paris. He has served as President of the European Society of Haematology (1960), the International Haematology Society (1976–1978) and of the Academy of Sciences of the Institute of France (1983).

Bernard's massive scientific contributions have been recognised by the award of honorary doctorates at universities in many parts of the world. His research has ranged from the demonstration of the neoplastic nature of leukaemia (1933–1937) to the formulation of methods of treatment, thirty years later. Bernard gave the first description of the use of high dosage radiotherapy in the treatment of Hodgkin disease (1932) and he was a founder of the science of geographic haematology.

Bernard is a prolific author, having written 14 textbooks and monographs on the subject of haematology. In the later stages of his career he played an important role in debates on ethical issues in medicine.

## Reference

Bernard J, Soulier J-P (1948) Sur une nouvelle variété de dystrophie thrombocytaire hémorragipare congénitale. Sem Hôp Paris 24: 3217–3223

# Binder, K.H.

Binder syndrome, or naso-maxillary dysostosis comprises flattening of the fronto-nasal angle, hypoplasia of the nose and digital shortening. There is uncertainty concerning the genetic background of this disorder.

Karl Heinz Binder was born in 1923 in Worms, Germany. He completed his education in Wiesbaden and was then conscripted into the army as a medical orderly. In January 1942 he was captured in the Libyan desert and spent the next 5 years as a prisoner-of-war in Egypt, Canada and England.

In 1947 Binder entered the University of Mainz in order to study medicine and dentistry. After graduation in 1952 he was appointed as school dental officer at Ludwigshafen, Rhine, and he remained in this post until 1976. From 1956 until his retirement in 1989 Binder was also in orthodontic practice. In 1962 he documented the facial features of 3 children with the condition which now bears his name.

Binder married in 1954 and has 2 children and 3 grandchildren. In his leisure time, he enjoys philosophic literature, the piano, swimming and tennis.

## Reference

Binder KH (1962) Dysostosis maxillo-nasalis, ein arhinencephaloider Missbildungskomplex. Dtsch Zahnärztl Z 17: 438–444

# Bixler, D.

Antley-Bixler syndrome, comprises craniostenosis, choanal atresia and radioulnar synostosis. Inheritance is probably autosomal recessive

David Bixler was born in Chicago, Illinois, on 7 January 1929. He received his professional training at Indiana University, obtaining a dental qualification and a doctorate and he subsequently became professor of medical genetics. He established a department of oral facial genetics in the dental school and served as professor and chairman until retirement in 1993. Bixler advocated the development of a dental curriculum that would use genetics to bring about a better understanding of medical problems and their solution. In order to carry out this plan he initiated a training programme for graduate dentists, in which there was a broad and thorough experience in clinical genetics as well as some research experience.

Bixler has a deep interest in syndromes of the cranio-facial complex especially in the genetics of clefting. His work has challenged the currently accepted multifactorial/threshold model in the etiology of this malformation and in collaboration with several investigators worldwide, Bixler has shown that some clefts are caused by a single gene located on chromosome 4.

## Reference

Antley RM, Bixler D (1975) Trapezoidocephaly, midface hypoplasia and cartilage abnormalities with multiple synostoses and skeletal fractures. Birth Defects Orig Art Ser XI(2): 397–401

# Böök, J.A.

Böök syndrome is an autosomal dominant disorder comprising premolar aplasia, hyperhidrosis and premature whitening of the hair. The acronymic title "PHC syndrome" is sometimes used.

Jan A. Böök was born in Malmö, Sweden in 1915. After obtaining scientific qualifications at the University of Lund, he travelled widely in Europe in order to gain additional educational experience, and in 1940 he was appointed as research assistant at the Institute of Genetics, Lund. Böök became director of the Lund medical genetics research unit in 1946 and 3 years later he received a licence to practise medicine. He then undertook an extensive genetic investigation of a population in northern Sweden, which led to the award of a doctorate from the University of Uppsala. After the conclusion of this project Böök spent 2 years at the University of Minnesota, USA, where he investigated the genetic basis of mental retardation. In 1951 he returned to the University of Uppsala where he was elevated to professorial rank in 1959.

Böök has been a consultant with WHO, UN and UNESCO and has served on several Swedish parliamentary committees concerned with medical genetics. He has published more than 160 medical articles and he has received numerous honours and awards.

## Reference

Böök JA (1950) Clinical and genetical studies of hypodontia. I. Premolar aplasia, hyperhidrosis and canities prematura; a new hereditary syndrome in man. Am J Med Genet 2: 240–263

# Brachmann, W.

Brachmann-de Lange syndrome comprises mental retardation, stunted stature, a characteristic facies and variable limb malformations.

W. Brachmann was born in 1890 or thereabouts and by 1913 he had qualified in medicine and obtained an appointment as a clinical assistant at the Children's Hospital in Dresden. While in this post he wrote an account of the clinical and autopsy features of an infant who died of pneumonia at the age of 19 days. The title of his article provided details of the child's manifestations: *A case of symmetrical monodactyly, representing ulnar deficiency, with symmetrical antecubital webbing and other abnormalities, (dwarfism, cervical ribs, hirsutism)*. Brachmann was primarily concerned with the limb defects, but the characteristic facies of the condition are also clearly apparent in the illustrations in his paper. Brachmann commented in his article that his investigations were interrupted by an urgent order to report for military service in the German army. He was subsequently killed in action in World War I and his portrait and curriculum vitae were destroyed during World War II; for these reasons available biographical information is scanty.

In 1933, de Lange of Amsterdam, (*see The Man Behind the Syndrome* p. 39) reported 2 female infants with mental retardation and an unusual facies, and after a further presentation in 1941, the condition became known as the "Cornelia de Lange syndrome". In a review in 1985, Opitz commented "Brachmann's paper is a classic of Western Medical iconography, deserving to be commemorated in the eponym 'Brachmann-de Lange syndrome'"; this conjoined eponym is now generally accepted.

## References

Brachmann W (1916) Ein Fall von symmetrischer Monodactylie durch Ulnadefekt, mit symmetrischer Flughautbildung in den Ellenbeugen, sowie anderen Abnormalitaten (Zwerghaftigkeit, Halsrippen, Behaarung). Jahrbuch Kinderheilkd 84: 225–235

Opitz J (1985) The Brachmann-de Lange syndrome. Am J Med Genet 22: 89–102

# Campanacci M.

Campanacci syndrome or osteofibrous dysplasia of the tibia and fibula manifests as antero-lateral bowing of the shins, with osteolytic and sclerotic changes in the bony cortices. The pathogenesis is unknown.

Mario Campanacci was born in Italy on 13 January 1932 and qualified in medicine at the University of Bologna in 1956. He undertook specialist training in orthopaedics traumatology and surgical pathology, and in 1958 joined the staff of the Rizzoli Institute of Orthopaedics, University of Bologna. Campanacci became director of the tumor pathology research laboratory in 1963 and was elevated to his present status of Departmental Chief and Chairman of the Postgraduate School in 1980.

In 1981, in collaboration with his colleague M. Laus, Campanacci published an account of 35 patients with the condition which now bears his name. In their article, Campanacci and Laus differentiated this entity from other forms of fibrous dysplasia and emphasised that the clinical course becomes static after puberty.

Campanacci has had an active academic career, with a special interest in bone and soft tissue tumors. He has served as President of the European Society for Musculo-Skeletal Oncology and published 7 books and more than 370 medical articles.

## Reference

Campanacci M, Laus M (1981) Osteofibrous dysplasia of the tibia and fibula. J Bone Jt Surg 63A(3): 367–375

# Catel, W.

Catel-Manzke syndrome comprises the Robin malformation sequence and an accessory ossicle on the index finger.

Werner Catel was born in Mannheim in 1894 and completed his training in paediatrics at the Children's Clinic, University of Leipzig, in 1926. He had considerable scientific skills and gained steady promotion, being elevated to professorial status in 1932 and to the post of clinic director in the following year. In 1947 Catel was appointed to the headship of a long-stay hospital for children with tuberculosis and in 1954 he became director of the Children's Clinic in Kiel. During World War II he was involved with the Nazis' euthanasia programmes and for this reason he was compelled to take early retirement in 1960, being replaced in this post by Hans-Rudolf Wiedemann. Catel continued as a medical author and in 1961 he published a brief report of the palato-digital malformation which became known as the Catel-Manzke syndrome *(see* Manzke, p. 215). His name is also associated with those of Schwartz and Jampel in the title of a syndrome which comprises myotonia, skeletal dysplasia, limitation of joint movements and blepharophimosis *(see The Man Behind the Syndrome* Jampel, p. 213).

In 1979, at the age of 85 years, Catel was co-editor of a textbook for paediatric nurses. He died in Kiel in 1981.

## Reference

Catel W (1961) Differentialdiagnose von Krankheitssymptomen bei Kindern und Jugendlichen, 3rd edn. Vol 1. G Thieme, Stuttgart, pp. 218–220

# Cohen, M. Michael, Jr.

Cohen syndrome comprises low birth weight, short stature, mental retardation, ophthalmological abnormalities and a characteristic facies. Inheritance is autosomal recessive.

M. Michael Cohen, Jr., was born in Cambridge, Massachusetts in 1937. He was educated at the University of Michigan, Tufts University, the University of Minnesota and Boston University. He holds a bachelor's degree in anthropology, a dental qualification, a master's degree in oral and maxillofacial pathology, a doctorate in anthropology and a certificate in international health. During his postgraduate training Cohen completed his fellowship in syndromology and medical genetics with Robert Gorlin. He became professor of oral and maxillofacial surgery and of paediatrics at the University of Washington in Seattle and in 1981 he moved to a chair at Dalhousie University in Canada. He teaches several subjects, including general pathology and a graduate seminar in health care in developing countries.

In 1973 Cohen and his colleagues described an autosomal recessive disorder comprising hypotonia and obesity with facial and ocular abnormalities, which is now known as the Cohen syndrome; over 100 patients had been reported by 1995.

He is the author of over 230 articles in the medical literature and author, co-author or editor of 11 books and 28 chapters. His main research interests concern children with multiple anomalies, with special focus on craniosynostosis, holoprosencephaly and overgrowth syndromes.

## Reference

Cohen MM Jr, Hall BD, Smith DW, Graham CB, Lampert KJ (1973) A new syndrome with hypotonia, obesity, mental deficiency and facial, oral, ocular and limb anomalies. J Pediatr 83: 280–284

# Currarino, G.

Currarino complex comprises anorectal, sacral and presacral anomalies. The pathogenesis is uncertain.

Guido Currarino was born in Levanto, Italy in 1920. He qualified in medicine at the University of Genoa in 1945 and, after postgraduate training in paediatrics with Professor G. De Toni, moved to the USA, where he developed an interest in paediatric radiology. Currarino obtained specialist experience with Professor F.N. Silverman in Cincinnati and with Professor E.D.B. Neuhauser in Boston. In 1955 he was appointed to the staff of the Childrens' Hospital in Cincinnati and in 1960 he moved to the New York Hospital, Cornell University. Currarino became chief radiologist at the Dallas Childrens' Medical Center in 1965, and after retirement he continued part-time work in that department. Currarino's many contributions to his speciality were rewarded in 1995 when he received the Gold Medal of the Society for Paediatric Radiology.

## Reference

Currarino G, Coln D, Votteler T (1981) Triad of anorectal, sacral and presacral anomalies. AJR 137: 395–398

# de Barsy, A-M.

de Barsy syndrome comprises cutis laxa, corneal clouding and mental retardation. Inheritance is autosomal recessive.

Anne-Marie de Barsy was born on 7 April 1939 at Lier, a small town between Antwerp and Brussels in Belgium. Her maiden name was de Cannart d'Hamale and she was the third of a family of 13 children. After education in Rymenam (Mechelen) and at a boarding school in Brussels, she studied medicine at the Catholic University of Louvain, qualifying in 1965. In the same year, she married her fellow student Thierry de Barsy, with whom she commenced neurological training under van Bogaert at the Bunge Institute in Berchem-Antwerp, Belgium.

In her second year of neurological training, de Barsy examined a child with stunted stature, delayed developmental milestones and a progeroid appearance. In collaboration with her paediatric colleagues, Moens and Dierckx a case report was published and de Barsy's name was subsequently applied to the disorder.

Since 1969, when she received her certification as a neurologist, de Barsy has worked at the Vezalius Hospital in Brasschaat near Antwerp, Belgium. She has recently developed an interest in palliative medicine and has undertaken training in France, England and Belgium. de Barsy's husband is professor of neurology at the Catholic University of Louvain; the couple have 2 daughters and 2 grandchildren.

## Reference

de Barsy AM, Moens E, Dierckx L (1968) Dwarfism, oligophrenia and degeneration of the elastic tissue in skin and cornea. A new syndrome? Helv Paediatr Acta 23(3): 305–313

# de la Chapelle, A.

de la Chapelle dysplasia, also known as atelosteogenesis type II, is a lethal form of neonatal dwarfism in which gross limb shortening is associated with a characteristic triangular configuration of the radius and ulna. Inheritance is autosomal recessive.

Albert de la Chapelle was born in 1933 in Helsinki, Finland and qualified in medicine in that city in 1957. After obtaining a doctorate and experience in internal medicine, he trained in biochemistry at the Columbia University, New York. In 1974 de la Chapelle was appointed as Professor and Chairman of Medical Genetics at the University of Helsinki and he became Chief Physician in Clinical Genetics in 1977; he still occupied these positions in 1995.

de la Chapelle has undertaken research in London, Paris and the USA, initially in human cytogenetics and latterly in the molecular genetics of hereditary disorders. He is the author of more than 300 medical articles, including an account published in 1972, concerning stillborn siblings with the condition which now bears his eponym.

## Reference

de la Chapelle A, Maroteaux P, Havu N, Granroth G (1972) Une rare dysplasie osseuse lethale de transmission recessive autosomique. Arch Fr Pédiatr 29: 759–770

# DiGeorge, A.

DiGeorge syndrome comprises abnormalities of the thymus, parathyroids and great vessels which result from defective embryonic development of the third and fourth pharyngeal pouches. Inheritance is probably autosomal dominant.

Angelo M DiGeorge was born on 15 April 1921 in Philadelphia, USA, and was educated at the South Philadelphia High School. He entered Temple University in 1936 with a competitive scholarship, joined the School of Medicine in 1939 and graduated with honours in 1946. DiGeorge served as a US army medical officer in Linz, Austria during the period 1947–1949, thereafter returning to Temple University School of Medicine for higher training in paediatrics. He completed his residency at St Christopher's Hospital for Children in 1952, obtained the MS degree and then became a postdoctoral fellow in endocrinology at the Jefferson Medical College.

In 1953 DiGeorge was appointed as instructor in the department of paediatrics, Temple University Medical School, where he spent the rest of his fruitful career. He received regular promotion and became professor of paediatrics in 1967; he retained this rank until his retirement in 1991, when he was granted emeritus status. Throughout his career, DiGeorge has been at the forefront of his speciality, and in 1990, at the age of 69 years, he received recertification in paediatrics and endocrinology from the American Board.

DiGeorge has been a member of numerous professional societies, committees and editorial boards. He has made notable contributions to undergraduate and postgraduate education, and in 1993 he was honoured with the establishment of the annual Angelo DiGeorge award for the best contribution to the instruction of the junior medical staff of his hospital.

## Reference

DiGeorge AM (1968) Congenital absence of the thymus and its immunologic consequences: concurrence with congenital hypoparathyroidism. Birth Defects Orig Art Ser IV(I): 116–121

# Eriksson, A.W.

Forsius-Eriksson syndrome or ocular albinism (*see* Forsius, p. 210).

Aldur Wictor Eriksson was born on 7 January 1927 at Geta, on the Åland Islands, Finland. He studied medicine at the University of Helsinki, qualifying in December 1956. At an early stage in his postgraduate career, Eriksson became interested in human genetics and in 1964 he obtained a post at the Finnish National Research Council for Medical Sciences. He became involved in the International Human Adaptability Biological Program, serving as chairman of the Finnish section; his work in this field included expeditions to study the Eskimos of North Western Greenland, the Icelanders, the Lapps and Skolts of northern Finland, the Volga-Finns of Mari, and the Komi of Russia.

In 1971 Eriksson was appointed as Professor of Human Genetics at the Free University of Amsterdam. He remained in this post until his retirement, when he moved back to Helsinki with emeritus status. Eriksson is the author or co-author of more than 500 medical publications, mainly concerning twin studies and the genetics of isolates and in 1986 his contributions were recognised by the award of the Gregor Mendel Medal. During 1986–89 he was elected president of the International Society for Twin Studies. Eriksson is married to May-Britt, formerly Helenius, an instructor in gymnastic dance and the couple have a son and 4 grandchildren. In 1995, during his retirement, Eriksson was occupied with his researches in population genetics and twinning and his hobbies of physical training and island lifestyles.

## Reference

Forsius H, Eriksson AW (1964) Ein neues Augensyndrom mit X-chromosomaler Transmission. Eine Sippe mit Fundusalbinismus, Foveahypoplasie, Nystagmus, Myopie, Astigmatismus und Dyschromatopsie. Klin Mbl Augenh 144(3): 447–457

# Fallot, E-L.A.

Fallot tetralogy is a structural cardiac abnormality comprising a ventricular septal defect, over-riding aorta, pulmonary stenosis and right ventricular hypertrophy.

Etienne-Louis Arthur Fallot was born on 29 September 1850 at Sète on the southern coast of France. After a brilliant scholastic career at the Lycée in Marseilles he became a medical student at the University of Montpellier in 1867. Following an internship at Marseilles, Fallot obtained a doctorate with a thesis on the subject of pneumothorax. He then became involved in the teaching of pathological anatomy and undertook his classical investigations of the anatomical basis of cyanotic cardiac disease. In 1888 Fallot published details of the pathological features of 2 persons with the condition which now bears his name, together with a detailed review of 50 other reports which he had culled from the literature. Although the tetralogy had previously been described, Fallot was the first to make clinico-pathological correlations in this disorder.

In 1888 Fallot was called to the chair of hygiene and legal medicine at the University of Marseilles. He occupied this post until his death on 30 April 1911; at his specific request, no obituary was written.

## Reference

Fallot E-L (1888) Contribution à l'anatomie pathologique de la maladie bleue (cynose cardiaque). Marseille Méd 25: 77–93; 138–158; 207–223; 341–354; 370–386; 403–420

# Filippi, G.

Filippi syndrome comprises an unusual facies, microcephaly, growth delay, mental retardation and syndactyly. This rare syndrome is an autosomal recessive trait.

Giorgio Filippi was born in Italy and studied medicine at the University of Rome, qualifying in 1959 and undertaking further training in paediatrics. In 1963 he joined the department of genetics of the Italian National Research Council and worked under the direction of Professor Siniscalco in population genetics in Sardinia, mainly in X-chromosomal mapping.

In 1967 Filippi commenced a research fellowship with Professor V.A. McKusick at the Johns Hopkins Hospital, Baltimore, USA and in 1969 he returned to the department of paediatrics, University of Rome, as lecturer in medical genetics. He was appointed as professor of medical genetics at the University of Trieste in 1973 and still occupied this post in 1995. Filippi's main academic interests involve clinical genetics, with special emphasis on the fragile-X syndrome.

## Reference

Filippi G (1985) Unusual facial appearance, microcephaly, growth and mental retardation and syndactyly. A new syndrome? Am J Med Genet 22: 821–824

# Forsius, H.

Forsius-Eriksson syndrome is a form of ocular albinism, which is characterised by defective vision and nystagmus. This X-linked disorder was delineated in the population of the Åland Islands in the Baltic Sea.

Henrik Forsius was born in Helsinki, Finland on 24 August 1921. He qualified in medicine at the University of Helsinki in 1948 and thereafter obtained a specialist qualification in ophthalmology. He received a doctorate in 1954 for a thesis on arcus senilis and in 1964 became professor of ophthalmology at the University of Oulu, Finland. Forsius remained in this post until 1986, when he took up an appointment as head of the population genetics department, Folkhälsan Institute of Genetics, Helsinki. In 1995, at the age of 73 years, Forsius was still active in this capacity.

Forsius has made many contributions in the field of hereditary eye disease, often in collaboration with his friend and colleague, Professor Aldur Eriksson (*see* p. 209). These include studies of X-linked retinoschisis, choroideremia, cornea plana congenita and granular corneal dystrophy. Forsius has travelled widely in order to investigate the effects of climatic conditions on the eye and he has undertaken investigations in the populations of Finland, Iceland, Canada, USA, Peru, Ecuador, Tunis, India, Cuba and Rwanda.

Forsius' wife, Harriet, is an associate professor of child psychiatry. The couple have 4 children.

## Reference

Forsius H, Eriksson AW (1964) Ein neues Augensyndrom mit X-chromosomaler Transmission. Klin Monatsblätter für Augenheilkunde 114: 447–457

# Fryns, J-P.

Fryns syndrome is a potentially lethal malformation complex comprising a coarse facies, corneal clouding, cleft soft palate, pulmonary hypoplasia, diaphragmatic defects and abnormalities of the distal portions of the limbs and the digits. Inheritance is autosomal recessive.

Jean-Pierre Fryns was born in Belgium in 1946 and studied medicine at the Catholic University of Leuven. He trained in paediatrics and in 1976 became head of the Clinical Genetics Unit at the Leuven University Hospital. His main academic interests are constitutional cytogenetics, genetic mental retardation, dysmorphology, the psychosocial aspects of mental handicap and the behavioural phenotype in mental retardation syndromes. He is the author or co-author of more than 950 articles and abstracts, the editor-in-chief of *Genetic Counselling (Journal de Génétique Humaine)* and a member of the editorial board of the *Journal of Medical Genetics*.

Fryns is the president of the Belgian group for the Scientific Study of Mental Deficiency and he is deeply involved in the training of medical and paramedical personnel in this field. Fryns married Gisela Everaerts in 1970 and the couple have 2 daughters.

## Reference

Fryns JP, Moerman F, Goddeeris P, Bossuyt C, Van den Berghe H (1979) A new lethal syndrome with cloudy corneae, diaphragmatic defects and distal limb deformities. Hum Genet 50: 65–70

# Fuhrmann, W.J.G. (1924–1995)

Fuhrmann syndrome comprises fibular hypoplasia, femoral bowing, digital abnormalities and other variable malformations. The condition is probably an autosomal recessive trait.

Walter Fuhrmann was born in Berlin in 1924 and received his education in that city. He was conscripted for military service in 1942 and after the armistice, studied medicine at the Humboldt and Freie Universities, Berlin qualifying in 1951. Following internships in New Jersey and Boston, USA, Fuhrmann returned to Germany and trained in paediatrics at the Kaiserin Auguste Viktoria Haus, Berlin. In 1963 he was a Fellow at the Children's Memorial Hospital, Northwestern University, Chicago, and from 1964 until 1967 he held a post at the Institute of Human Genetics, University of Heidelberg. Fuhrmann succeeded to the Chair of Human Genetics at the Justus-Liebig University, Giessen in 1967, where he remained until he received emeritus status in 1992. In 1980 he documented 3 siblings with the condition which now bears his name; the parents were Turkish-Arabian members of a Christian minority, working in Germany. In a subsequent article, Fuhrmann and his colleagues reported the successful antenatal diagnosis of a further affected sibling in the family. Fuhrmann had an active academic career and was the author of a monograph and more than 200 medical articles. He died suddenly in his home town, Giessen, on October 19th 1995.

## Reference

Fuhrmann W, Fuhrmann-Rieger A, de Sousa F (1980) Poly-, syn- and oligodactyly, aplasia or hypoplasia of fibula, hypoplasia of pelvis and bowing of femora in three sibs – a new autosomal recessive syndrome. Eur J Pediatr 133: 123–129

# Hecht, F.

Hecht-Beals syndrome comprises trismus and pseudocamptodactyly. Inheritance is autosomal dominant.

Frederick Hecht was born in 1930 in Baltimore, Maryland, USA, the second of 3 sons of Malcolm Hecht, a department-store merchant, and Lucile Burger Levy, from Louisville, Kentucky. Raised in Pikesville, Maryland, he graduated from Baltimore Friends School in 1948. He received his BA degree with honours in French from Dartmouth College in 1952 and, after graduate studies, served as interpreter-translator in Russian and German in the US Army. Following discharge in 1955 he attended Boston University and continued his education at the University of Rochester School of Medicine and the Strong Memorial Hospital, where he obtained his MD with distinction in 1960 and served his internship and residency in paediatrics. At the University of Washington in Seattle from 1962 to 1965, Hecht was a postdoctoral research fellow and assistant in medical genetics with Dr Arno Motulsky and an instructor in medical genetics and paediatrics. From 1965 to 1978 Hecht was co-director of the genetics clinic at the University of Oregon Medical School (Oregon Health Sciences University) in Portland.

With his wife, Dr Barbara K. Hecht, Hecht founded the Southwest Biomedical Research Institute in 1978 in Arizona and served as president and director of the Genetics Center and the Cancer Center of SBRI until 1989. In 1992 Hecht became professeur associé at the faculty of medicine at the University of Nice in France. He is co-author of the monograph *Fragile Sites on Human Chromosomes* and he edited *Trends and Teaching in Medical Genetics*. Hecht's interests outside of medicine include languages, literature and writing and his large family of 6 children and 6 grandchildren.

## Reference

Hecht F, Beals RK (1969) Inability to open the mouth fully: An autosomal dominant phenotype with facultative camptodactyly and short stature. Birth Defects Original Article Series 5(3): 96–98

# Hecht, J.T.

Hecht-Scott syndrome comprises terminal transverse defects of the limbs and congenital cardiac malformations. The mode of inheritance is uncertain, although autosomal recessive transmission seems likely.

Jacqueline Hecht graduated from New York University with a BSc degree in Biology and subsequently obtained a Masters degree in Genetics and Genetic Counselling at the University of Colorado Health Sciences Centre, Denver. She then proceeded to obtain a PhD in genetic epidemiology at the University of Texas School of Public Health. Hecht's academic contributions have included the delineation of the natural history of genetic disorders and the identification of the molecular defects in skeletal dysplasias and craniofacial conditions.

In 1995, Hecht was an Associate Professor at the University of Texas Health Centre, Houston. She has the distinction of being one of the very few non-medical scientists whose name has been attached to a genetic syndrome.

## Reference

Hecht JT, Scott CI Jr (1981) Limb deficiency syndrome in half-sibs. Clin Genet 20: 432–437

# Herrmann, C. Jr.

Herrmann syndrome comprises diabetes mellitus, nephropathy, photomyoclonus, deafness and dementia. Inheritance is probably autosomal dominant.

Christian Herrmann Jr. was born on 25 January 1921, in Lansing, Michigan, USA. He studied medicine at the University of Michigan Medical School, qualifying in 1944. After internship and residency in Detroit, Herrmann served from 1946 until 1948 as a medical officer at the United States Naval Hospital, San Diego, California. He then trained in neurology at the Neurological Institute of Columbia Presbyterian Medical Center, New York receiving board certification in his speciality in 1952 and thereafter obtaining a post as instructor in neurology. Herrmann moved to the University of California School of Medicine, Los Angeles in 1954, as assistant professor of neurology, succeeding to professor in 1969 and vice chairman in 1970. He held this rank until his retirement in 1986, when he was elevated to emeritus status.

Herrmann's main academic focus has been myasthenia gravis and during his career he has studied more than 950 persons with this disorder. He has published numerous articles on this subject and has held office in various lay and professional bodies which are concerned with the condition. His endeavours in this field have been recognised by the bestowal of several honours and awards.

## Reference

Herrmann C Jr., Aguilar MJ, Sacks OW (1964) Hereditary photomyoclonus associated with diabetes mellitus, deafness, nephropathy, and cerebral dysfunction. Neurology 14: 212–221

# Herrmann, J.

Herrmann facio-audio-symphalangism syndrome comprises a broad nose, conductive deafness and synostosis of the digits. Inheritance is autosomal dominant.

Jürgen Herrmann was born in 1941. He spent his childhood in post-World War II Northern Germany and qualified in medicine at the University of Hamburg in 1966. A lecture by Professor Widukind Lenz aroused his interest in medical genetics and on Lenz's recommendation, he commenced a 1-year fellowship with Opitz at Madison, Wisconsin, USA in 1967, concurrently with Hans Dieker (*see* p. 55). Herrmann enjoyed the intellectual stimulation of working with Opitz and he remained in Wisconsin for the next 12 years. During this period he was involved in a programme of syndromic delineation and his collaboration with Opitz's colleague, Pallister, culminated in 1969 in his marriage to the latter's daughter.

Herrmann remained in Madison until 1978 when he moved to the Medical College of Wisconsin at Milwaukee, 70 miles to the east. In 1982 he entered private practice and his group now provides clinical, prenatal and laboratory services to the community.

Herrmann and his wife, Andrea, have 3 children. They share a liking for travel, tennis and skiing and they have developed an interest in contemporary studio glassware.

## Reference

Herrmann J (1974) Symphalangism and brachydactyly syndrome: report of the WL symphalangism-brachydactyly syndrome: review of literature and classification. Birth Defects Orig Art Ser X(5): 23–53

# Hirschhorn, K.

Wolf-Hirschhorn syndrome (*see* Wolf, p. 222)

Kurt Hirschhorn was born on 18 May 1926 in Vienna, Austria. He received his early education in Vienna and attended high school in Pittsburgh, USA. After 3 years in the US Army Hirschhorn continued with his studies, obtaining his BA cum laude in 1950 from New York University and the MD degree in 1954 from New York University School of Medicine. He was elected to both Phi Beta Kappa and Alpha Omega Alpha.

Hirschhorn served his residency in internal medicine at Bellevue Hospital, New York, remained there as a fellow in metabolic diseases and subsequently undertook a fellowship in human genetics in Uppsala, Sweden. In 1958 he was appointed to the staff of the New York University school of medicine and in 1966 became professor of paediatrics at Mount Sinai School of Medicine in New York, where he established a new medical genetics programme. From 1968 to 1976 Hirschhorn was the Arthur J. and Nellie Z. Cohen professor of genetics at Mount Sinai School of Medicine. In 1977 he became the Herbert H. Lehman professor and chairman of the department of paediatrics at Mount Sinai School of Medicine and paediatrician-in-chief at Mount Sinai Hospital, positions which he still held in 1994.

Hirschhorn's work in genetics began with studies of hypercholesterolaemia and other lipid disorders and, while in Sweden, he entered the fields of tissue culture genetics and cytogenetics. At Mount Sinai he became involved with various aspects of biochemical genetics and, more recently, in molecular genetics. In the early 1960s he discovered the mixed lymphocyte reaction which was fundamental to the fields of cellular immunology and immunogenetics. He also made major contributions to the understanding of several inborn errors, including lysosomal enzyme defects and Menkes syndrome. Hirschhorn's longstanding interest and leadership in medical genetics has resulted in the publication of more than 350 articles and book chapters and he has been the co-editor of *Advances in Human Genetics* for several years.

Hirschhorn is married to Rochelle Hirschhorn MD, also a major contributor to the field of human genetics. The couple have 3 children, Joel and Lisa who are also medically qualified, and Melanie, an attorney.

## Reference

Hirschhorn K, Cooper HL (1961) Apparent deletion of short arms of one chromosome (4 or 5) in a child with defects of midline fusion. Human Chromosome Newsl 4: 14

# Ito, M.

Ito hypomalanosis comprises skin whorls and streaks together with variable abnormalities in the brain, eyes and skeleton. Chromosomal mosaicism is present in a proportion of affected persons.

Minoru Ito was born in 1892 in Tokyo, Japan. He qualified in medicine at the Tokyo Imperial University in 1918 and commenced training in dermatology under Professor K. Dohi. In 1924 he was appointed to the post of associate professor of dermatology at Kanazawa Medical School and he later became full professor and departmental chairman. In 1937 Ito was elected as Professor of Dermatology and departmental chairman at Tohaku University. Towards the end of World War II, the environs of his hospital were destroyed by repeated allied bombardments and he overcame many difficulties at this time.

Ito was a popular lecturer, with a wide repertoire. He trained a total of 84 dermatologists, of whom 9 eventually became professors at various universities in Japan. Ito represented his country at the 1952 International Congress of Dermatology, London and was subsequently elected as president of the Japanese Dermatological Association and the Japanese Association of Allergology. He was also an honorary member of the Japanese Dermatological Association and the Society of Investigative Dermatology.

After his retirement from Tohaku University in 1957, Ito became director of the Aomori Central Hospital. He finally retired in 1968, moving to Kamakura, where he enjoyed classical Kabuki theatre. Ito died in Kamakura in 1986, at the age of 94 years.

## Reference

Ito M (1952) Studies on melanin. XI. Incontinentia pigmenti achromians. A singular case of nevus depigmentosus systematicus bilateralis. Tohoku J Exp Med 55(suppl): 57–59

# Ivemark, B. I.

Ivemark asplenia syndrome is a combination of situs inversus, agenesis of the spleen, and congenital heart defects; a variant is Ivemark polysplenia, in which multiple spleens are associated with less severe heart defects. Inheritance of both conditions is probably autosomal recessive.

Biörn I Ivemark was born in Karlstad, Sweden, on 4 May 1925. After qualifying in medicine at the Karolinska Institute, Stockholm he received a Rockefeller grant to study paediatric pathology with Farber at the Children's Hospital, Boston, Massachusetts. During his stay in Boston, Ivemark collected 14 new cases of asplenia; he published his doctoral thesis on the subject in 1955, and was awarded a PhD in pathology.

In collaboration with John Hellström, Ivemark published a monograph on primary hyperparathyroidism in 1962. Ivemark also wrote a textbook of paediatric pathology in 1971; a German edition appeared in 1974. Ivemark was professor of paediatric pathology at the Karolinska Institute from 1960 until 1983, where he performed diagnostic and teaching services, and he was twice elected as "best teacher" by a majority of students. He co-founded the Nordic Society of Paediatric Pathology in 1964 with Lars Hjelt of Helsinki and was chairman until 1982. Ivemark was also president of the Swedish Association of Pathologists from 1977 until 1981.

Ivemark retired in 1983, and in 1995 he was living in the south of France together with his wife, Carolina, a gastroenterologist and coagulationist and their 2 children. He is an amateur historian with an interest in WWII, Napoleon I, and Charles XII. He is also a dedicated gastronomer and chef and and in 1980 his talents led to his promotion from Chevalier to Officier de la Chaîne des Rôtisseurs. Ivemark cooks on a regular basis for family and friends from many countries.

## Reference

Ivemark BI (1955) Implications of agenesis of the spleen on the pathogenesis of cono-truncus anomalies in childhood. An analysis of the heart malformations in the splenic agenesis syndrome. With fourteen new cases. Acta Paediatrica Scand. 44, suppl 104: 1–110

# Kaufman, R. L.

Kaufman oculocerebrofacial syndrome comprises mental retardation, ocular abnormalities, high palate and micrognathia. The disorder is an autosomal recessive trait.

Robert L. Kaufman was born in St Louis, Missouri on 1 November 1937. He studied at Washington University, Missouri, receiving an undergraduate degree in chemistry in 1959 and his medical qualification in 1963. During the senior year of medical school he undertook a clerkship in medical genetics with Victor A. McKusick at the Johns Hopkins University School of Medicine, Baltimore.

After training in internal medicine at the Vanderbilt University Hospital, Kaufman spent 2 years as staff associate at the National Institute of Arthritis and Metabolic diseases, National Institutes of Health, Bethesda, Maryland. In 1968 he returned as chief resident at the Washington University Medical Service at St Louis City Hospital. A medical genetics fellowship was subsequently undertaken at Washington University Medical Center and he became director of the Genetics Clinic at St Louis Children's Hospital in 1970.

Kaufman is a Diplomate of the American Board of Medical Genetics and a Fellow of the American College of Physicians. In 1994 he was engaged in the practice of internal medicine in the St Louis area.

## Reference

Kaufman RL, Rimoin DL, Prensky AL, Sly WS (1971) An oculocerebrofacial syndrome. Birth Defects Original Article Series VII(1): 135–138

# Kearns, T.P.

Kearns-Sayre[1] syndrome is a heterogeneous disorder comprising progressive external ophthalmoplegia, retinitis pigmentosa and cardiomyopathy. Onset occurs in childhood and additional endocrinological and neurological abnormalities may develop. Mitochondrial inheritance is likely.

Thomas P. Kearns was born in 1922 in Louisville, Kentucky, where his father was a railroad worker. He had a happy, although impoverished childhood, but with the encouragement of an uncle who was a physician he succeeded in qualifying in medicine at the University of Louisville in 1946. After internship he served as a medical officer in the United States Army and in 1949 he commenced residency training in ophthalmology at the Mayo Clinic, Rochester. Kearns subsequently joined the clinic staff and progressed to become full professor at the Mayo Medical School. His major academic interests are in the field of neuro-ophthalmology and amongst other contributions, he published the first paper on the use of CT scanning in ophthalmology.

He married Mary Lou Trigg in 1944 and the couple have a daughter and a son. Kearns retired from practice in 1987 and now divides his time between his homes in Rochester and Louisville.

## Reference

Kearns TP, Sayre GP (1958) Retinitis pigmentosa, external ophthalmoplegia and complete heart block. Arch Ophth 60: 280–289

# Killian, W.

Pallister-Killian syndrome comprises mental retardation, multiple congenital anomalies, short limbs and a coarse facies. The condition results from isochromosome 12p mosaicism.

Wolfgang Killian was born on 8 May 1944 in Korneuburg, Austria. He qualified in medicine at the University of Vienna in 1971 and after internships, undertook training in clinical genetics at the Institute of Human Genetics, Marburg, Germany. In 1975 Killian obtained an appointment under Professor A Rett at the Department for Handicapped Children, Vienna; this post provided the opportunity for further study and in 1980 he received a degree in physical anthropology. In the following year he investigated an infant with severe mental retardation and an unusual distribution of hair. Publication of a report and the recognition of other similar case descriptions in the literature led to the use of the conjoined eponym "Pallister-Killian syndrome". In 1992 Killian was a consultant in syndromic diagnosis and genetic counselling for paediatric institutions in Vienna, based at the St Anna Children's Hospital (*see* Pallister, p. 217).

## Reference

Killian W, Zonana J, Schroer R (1983) Abnormal hair, craniofacial dysmorphism and severe mental retardation – a new syndrome? J Clin Dysmorphol 1: 6–13

# Kniest, W.

Kniest dysplasia is characterised by dwarfism, a flat facies, myopia and bulky joints. Inheritance is autosomal dominant and most affected persons represent new gene mutations.

Wilhelm Kniest was born on 25 October 1919 in Obersuhl-Hessen, Germany, where his father was in business. He studied medicine at the universities of Berlin, Leipzig, Vienna and Würzburg, qualifying in Göttingen in 1946. He then trained in paediatrics under Professor J. Ibrahim at the Children's Hospital, Jena, where he successfully combined his clinical and scientific activities. Kniest's early investigations were in the field of tuberculosis and he became interested in the radiodiagnosis of this disorder. He then focussed on the genetic skeletal dysplasias, and in 1952 he documented the disorder which now bears his name. After the death of Ibrahim in 1953, Kniest became head of the newly established children's hospital in Naumburg-Saale, where he spent the remainder of his career.

## Reference

Kniest W (1952) Zur Abgrenzung der dysostosis enchondralis von der Chondrodystrophie. Z Kinderheilk 70: 633–640

---

[1]Dr G.P. Sayre was a neuropathologist at the Mayo Clinic. He died in 1992.

# Kuroki, Y.

Niikawa-Kuroki syndrome (*see* Niikawa, p. 217).

Yoshikazu Kuroki was born in 1937 in Kagoshima, Japan. He graduated from the Faculty of Medicine, Kyushu University in 1963 and trained in paediatrics at Kyushu University, and in cytogenetics at Hokkaido University under Professor S. Makino, a pioneer in this field. From 1969 to 1973 Kuroki served as a paediatrician in the National Fukuoka Central Hospital. He moved to the Kanagawa Children's Medical Centre in 1974 becoming professor of medical genetics in 1976. Thereafter, he established the first population-based birth defects monitoring system in Japan. In 1982 Kuroki became director of the Division of Medical Genetics, Kanagawa Children's Medical Centre and since 1987 he has also been professor of paediatrics at the St Marianna University School of Medicine.

Kuroki is the author of numerous monographs and scientific papers on clinical cytogenetics and congenital malformation syndromes, and he is currently editor of the *Japanese Journal of Human Genetics*, and of *Congenital Anomalies*. Kuroki's professional affiliations include the Japan Society of Human Genetics, the Japanese Teratology Society, Societas Paediatrica Japonica, and the European Society of Human Genetics. He is also a Member of the Order of the International Fellowship. In his spare time, Kuroki enjoys ornithology and travel.

## Reference

Kuroki Y, Suzuki Y, Chyo H, Hata A, Matsui I (1981) A new malformation syndrome of long palpebral fissures, large ears, depressed nasal tip, and skeletal anomalies associated with postnatal dwarfism and mental retardation. J Pediatr 99: 570–573

# Labhart, A.

Prader-Labhart-Willi syndrome comprises obesity, stunted stature, small hands and feet and mental retardation. An interstitial deletion in the long arm of chromosome 15 is present in a proportion of affected persons.

Alexis Labhart was born on 4 May 1916 in St Petersburg, Russia, of Swiss parents. His family returned to Switzerland in 1918 and he received his schooling in Basle. Labhart entered the medical faculty of the University of Basle in 1938 and qualified in 1944, writing a thesis on *Tuberculosis in Concentration Camps*. He pursued his interest in tuberculosis at a sanatorium in Davos and in 1947 became resident in internal medicine at the Insel Hospital, Berne. In 1951 Labhart was a research fellow in medicine at the Harvard Medical School and the Peter Bent Brigham Hospital in Boston and thereafter took up a post at the University of Zurich. Here he gained steady promotion, becoming head of the metabolic unit in 1961 and full professor of internal medicine in 1969.

Labhart is the co-author of a textbook entitled *Clinical Endocrinology* which went into second editions in both German and English. He has also written 185 medical articles, mainly in the field of endocrinology, metabolic disorders and diabetes.

(*see* Willi, p. 195; *The Man Behind the Syndrome* Prader, p. 223.)

## Reference

Prader A, Labhart A, Willi H (1956) Ein Syndrom von Adipositas, Kleinwuchs, Kryptorchismus und Oligophrenie nach myatonieartigem Zustand im Neugeborenenalter. Schweiz Med Wschr 86: 1260–1261

# Leigh, D.

Leigh syndrome, or subacute necrotising encephalomyelopathy, is a progressive, potentially lethal neurological disorder of infancy characterised by specific histopathological changes in the brain. Pyruvate carboxylase deficiency, inherited as an autosomal recessive trait, has been demonstrated in some affected persons.

Denis Leigh was born in 1915 in Manchester, England, and attended school and university in that city. He qualified in medicine shortly before the onset of World War II and, while serving as house surgeon to Geoffrey Jefferson, he assisted in the first operation for clipping an intracerebral aneurysm. During his military service Leigh worked with a team of neurosurgeons at the military hospital for head injuries, Oxford, and with mobile neurosurgical units in the field, serving in India and Burma. After the war Leigh gained further experience at the London and Maudsley hospitals, before undertaking a Nuffield fellowship in psychiatry at the Massachusetts General Hospital, Boston. He then returned to the Maudsley Hospital, London, where he held an appointment as consultant psychiatrist for the remainder of his career. He retained his interest in neuropathology and in 1951, in the laboratory of the Institute of Psychiatry, he documented the clinical and histopathological features of the condition which bears his name.

In 1994 Leigh was in full-time medico-legal practice, living in a village near Sevenoaks, Kent, England.

## Reference

Leigh D (1951) Subacute necrotizing encephalomyelopathy in an infant. J Neurol 14: 216–221

# Lemli, L.

Smith-Lemli-Opitz syndrome is an autosomal recessive disorder in which mental and developmental delay are associated with characteristic minor facial and digital abnormalities.

Luc Lemli was born on 26 December 1935 in Aalst, Belgium. He was educated at St Jozef's College, Aalst and studied medicine at the University of Louvain, qualifying magna cum laude in 1960. After internship, Lemli was a resident in paediatrics at the University of Wisconsin, Madison, USA (1960–62) and he then became an NIH Fellow in paediatric endocrinology at the same centre. In 1964 Lemli returned to Belgium, where he was a paediatric resident at the Stedelijk Ziekenhuis, Dendermonde (1964–65) and obtained his board qualification in paediatrics. He combined hospital and private practice, becoming head of paediatrics in 1976 at the OLV Troost Hospital, Dendermonde. In 1994 Lemli retired from hospital service but retained his busy private practice. Lemli published a number of articles in the field of morphology and paediatric endocrinology. He is married to Nicole Cardon and has 2 children and 4 grandchildren.

(*See The Man Behind the Syndrome* Smith, p. 163; Opitz, p. 223.)

## Reference

Smith DW, Lemli L, Opitz JM (1964) A newly recognized syndrome of multiple congenital anomalies. J Pediatr 64: 210–217

# Leroy, J.G.

Leroy syndrome, inclusion-cell disease (I-cell disease) or mucolipidosis II, is an autosomal recessive oligosaccharidosis, which is characterised by stunted stature, psychomotor retardation, coarse facies, rigid joints and skeletal deformities.

Jules Leroy was born in Belgium in 1934 and qualified in medicine and in biochemistry at the Ghent University School of Medicine. He then became a fellow in paediatrics at the Children's Hospital, Boston, and spent 3 years in medical genetics and paediatrics at the University of Wisconsin, USA, where he obtained a PhD in medical genetics. Leroy was professor of genetics and medical genetics at Antwerp University medical school from 1967 to 1983, and he was then appointed to his current position of professor and chairman of paediatrics at the University of Ghent. Leroy has published many scientific papers relating to the oligosaccharidoses and other metabolic disorders and in the fields of paediatrics, genetics and clinical dysmorphology.

## Reference

Leroy JG, DeMars RI, Opitz JM (1969) I-cell disease. Birth Defects: Original Article Series V(4): 174–185

# Lynch, H.T.

Lynch syndrome I represents a familial proclivity to nonpolyposis colorectal cancer. Lynch syndrome II, hereditary nonpolyposis colorectal cancer (HNPCC), has the additional feature of a very high risk for other adenocarcinomas. Both disorders are inherited as autosomal dominant traits.

Henry T. Lynch was born in Lawrence, Massachusetts on 4 January 1928 and married Jane Frances Smith in 1951. The couple have a son and 2 daughters. Lynch studied at the Universities of Oklahoma and Denver before embarking upon research in human genetics at the University of Texas, Austin. He then shifted his interest to medicine and in 1960 he graduated from the University of Texas, Galveston. He completed residency training in internal medicine at the University of Nebraska College of Medicine, Omaha, and gained further experience through a fellowship in clinical oncology. In 1967 he joined the faculty of the Creighton University School of Medicine, Omaha, Nebraska as professor and chairman of the department of preventive medicine. In 1980 he became professor of medicine in the same institution, and director of the Cancer Center in 1994.

The main thrust of Lynch's work has been cancer genetics and he has published more than 400 articles on this subject. He is also the author of 12 books, some of which include his wife and/or son as co-authors. Lynch has served on several national committees which are concerned with the prevention of cancer and he is recognised as a pioneer in this field.

## Reference

Lynch HT, Krush AJ (1967) Heredity and adeno-carcinoma of the colon. Gastroenterology 53: 517–526

# Mainzer, F.

Saldino-Mainzer syndrome comprises retinitis pigmentosa, nephropathy, cerebellar ataxia, stubby digits and stunted stature. Inheritance is probably autosomal recessive.

Frank Mainzer was born in Holland in 1939 and emigrated to the United States in 1949. He qualified in medicine at New York University School of Medicine and trained in radiology at the University of California, San Francisco. Since 1971 he has been chairman of the department of radiology at the St Francis Memorial Hospital and he also has the status of clinical professor of radiology at the University of California, San Francisco.

In the early stages of his career, at the time that he was a co-author of the article describing the condition which bears his name, Mainzer had an interest in paediatric skeletal radiology. In 1995 he was a general radiologist, with special expertise in magnetic resonance imaging (*see The Man Behind the Syndrome* Saldino, p. 226).

## Reference

Mainzer F, Saldino RM, Ozonoff MB, Minagi H (1970) Familial nephropathy associated with retinitis pigmentosa, cerebellar ataxia and skeletal abnormalities. Am J Med 49: 556–562

# Manzke, H.

Catel-Manzke syndrome (*see* Catel, p. 208).

Hermann Manzke was born in Stettin in 1933. He studied medicine at the Universities of Kiel and Freiburg, together with his monozygous twin brother, Eberhard, who subsequently became an internist and radiologist. Manzke qualified in 1959 and after residency training he joined the staff of the University Hospital for Children, Kiel, where he remained until 1986. He was accorded professorial status in 1975 and in 1987 he was appointed as medical director of the Kaiser Friedrich Children's Hospital at Norderney on the north sea coast of Germany.

Manzke's primary research interests have been in the fields of perinatology, respiratory physiology and clinical genetics. He has published more than 130 medical articles and he is the author of two monographs on infantile development. In 1963, while working under Professor H. R. Wiedemann at the Children's Hospital, Kiel, he reviewed the literature on chondrodysplasia punctata and drew attention to the frequent occurrence of cataracts and ichthyosis in affected persons. Manzke's name is sometimes associated with this form of the disorder, although his eponym is more often used in conjunction with that of Catel in the title of a syndrome which comprises malformations of the lower jaw and fifth fingers.

## Reference

Manzke H (1966) Symmetrische Hyperphalangie des zweiten Fingers durch ein akzessorisches Metacarpale. Röntgenfortschritte 105: 425–427

# Marshall, R.E.

Marshall-Smith syndrome comprises motor and mental retardation, accelerated skeletal maturation and an unusual facies. The mode of inheritance is uncertain.

Richard E. Marshall was born on 30 July 1933 in Long Beach, New York, USA. After schooling in New York he graduated at the Wesleyan University and in 1962 he obtained a medical qualification at Yale University School of Medicine. After completing his internship Marshall undertook research in developmental biology at the National Institute of Health. He then decided upon a career in paediatrics and in 1967 he commenced training in that speciality at the University of Washington. During this period Marshall worked with the late Professor David Smith, and together they delineated the disorder which now bears their conjoined eponym and co-edited a book entitled *Introduction to Clinical Pediatrics*. Marshall became interested in neonatology and in 1971 he was appointed as director of the neonatal unit at the St Louis Children's Hospital, where he remained until 1986. He was then appointed consecutively to professorships of neonatology at the Cleveland Metropolitan Hospital, Michigan State University and at the medical school, New Jersey.

Marshall was married in 1958 to Judy Moftey, a practicing attorney, and the couple have 2 daughters. His personal interests include reading, listening to classical music, art appreciation, golf and tennis (*see The Man Behind the Syndrome* Smith, p. 163).

## Reference

Marshall RE, Graham CB, Scott CR, Smith DW (1971) Syndrome of accelerated skeletal maturation and relative failure to thrive: a newly recognized clinical growth disorder. J Pediatr 78: 95–101

# Martinelli, B.

Acromesomelic dysplasia, Campailla and Martinelli form, is characterized by severe stunting of stature and shortening of the tubular bones of the forearms, shins, hands and feet. Inheritance is autosomal recessive.

Bruno Martinelli was born in 1937 in Venice, Italy. He qualified in medicine in 1961 and subsequently trained in orthopaedics and traumatology at the University of Padua. He then worked in the orthopaedic departments of the Universities of Padua, Sassari and Trieste and in 1988 he was appointed to the second chair of clinical orthopaedics at Trieste University. In addition to the management of trauma, his department is a centre for sports injuries and arthroscopy. Martinelli's long-standing interest in the osteochondrodysplasias commenced in 1970 when he was a co-author of an article documenting the condition which now bears his eponym. Outside his work Martinelli enjoys skiing, golf and his Ferrari car (*see The Man Behind the Syndrome* Campailla, p. 207).

## Reference

Campailla E, Martinelli B (1970) Deficit staturale con micromesomelia. Minerva Ortopedica 20: 1–5

# Mietens, C.

Mietens syndrome comprises mental retardation, small stature, corneal opacity, a narrow nose and dislocation of the radial heads. Inheritance is probably autosomal recessive.

Carl Mietens, the son of Carl Mietens Sr, a physician, was born in Edenkoben, Pfalz, Germany on 10 October 1933. He studied medicine at Heidelberg, Innsbruck, Hamburg and Munich and after qualification in 1959 undertook an internship in Atlantic City, USA. Mietens then held a research fellowship at the Childrens' Hospital, Philadelphia, before returning to Germany as senior paediatric resident at the Pediatric Clinic of the University of Würzburg. After obtaining a higher qualification in Paediatrics and Clinical Virology, he was promoted to the rank of associate professor in 1976. Thereafter, he became Director of the Children's Hospital of Bochum and Chairman of Paediatrics at the University of Bochum.

Mietens has had an active academic career and he has written more than 100 medical articles, including a description of 4 siblings with the disorder which bears his eponym.

## Reference

Mietens C, Weber H (1966) A syndrome characterized by corneal opacity, nystagmus, flexion contracture of the elbows, growth failure, and mental retardation. J Pediatr 69: 624–629

# Miller, J.Q.

Miller-Dieker lissencephaly syndrome (*see* Dieker, p. 55).

James Q. Miller, born 1926, studied at Haverford College and Columbia University College of Physicians and Surgeons. After qualification he completed neurological residency training at the University of Virginia followed by a 2-year fellowship in neuropathology at Harvard. While working at the Boston Children's Hospital, Miller documented two siblings with the lethal syndrome of lissencephaly and microcephaly, which subsequently bore his eponym.

In 1962 Miller joined the faculty of the University of Virginia School of Medicine at Charlottesville and established a cytogenetics laboratory in the days when pinto bean extract was the source of the mitogenic agent. He spent his career at the University and in 1995 he was alumni professor in neurology. His recent interests are congenital brain malformations and the management of neurological patients, particularly those with multiple sclerosis and epilepsy. Miller's professional activities include the teaching of predoctoral and postdoctoral students and he utilizes simulated patients to teach selected aspects of neurological history and examination.

Miller was assistant dean of the school of medicine for 6 years and chairman of the council on medical education for 8 years. He was chairman of the undergraduate education subcommittee and the section of education of The American Academy of Neurology and in 1995 he received the A.B. Baker award for leadership in American neurological education.

Outside his academic activities, Miller is an elder of the Westminster Presbyterian Church. His hobbies include gardening, hiking, camping and woodworking.

## Reference

Miller JQ (1963) Lissencephaly in two siblings. Neurology 13: 841–850

# Nievergelt, K.

Nievergelt syndrome is an autosomal dominant skeletal dysplasia characterised by stunted stature and predominant abnormalities in the forearms and lower legs.

Kurt Nievergelt was born in Zurich, Switzerland on 25 June 1913 and educated in that city. He then studied medicine in Zurich and Berlin, qualifying in 1938. After resident posts in pathology and bacteriology, Nievergelt trained in Orthopaedics and General Surgery. During this period he documented an affected father and 3 affected sons, all by different mothers, with the condition which now bears his name. In 1950, Nievergelt obtained a consultant appointment at the Balgrist Clinic, Zurich. He subsequently undertook further studies in Leiden, Paris and London, before entering private practice, where he remained until his retirement in 1979.

## Reference

Nievergelt K (1944) Positiver Vaterschaftsnachweis aufgrund erblicher Missbildungen der Extremitaeten. Arch Klaus Stift Vererbungsforsch 19: 157

# Niikawa, N.

Niikawa-Kuroki syndrome or Kabuki Make-Up syndrome comprises stunted stature, mental retardation and a characteristic facial appearance which resembles that assumed by Japanese traditional actors. The genetic background is uncertain but it is likely that affected persons represent new mutations for an autosomal dominant gene.

Norio Niikawa was born in Japan in 1942 and qualified in medicine at Hokkaido University in 1967. After his internship he trained in paediatrics from 1968 to 1971 at the Hokkaido University Hospital. Niikawa was then employed as a paediatrician in a small municipal hospital and thereafter he spent the period 1972 to 1975 in the embryology and cytogenetics laboratory, Department of Gynaecology and Obstetrics, Cantonal Hospital, Geneva, Switzerland. Here, under the direction of Professor Kajii, he demonstrated that dispermy and androgenesis were the mechanisms for human triploidy and hydatidiform mole, respectively.

Niikawa was an instructor of paediatrics at Hokkaido University School of Medicine from 1976 to 1984, and received his PhD in 1979. During this period he saw 5 patients who had a similar clinical picture, one of whom he had seen in infancy in the municipal hospital where he had previously worked. The disorder in these 5 patients were reported as "Kabuki make-up syndrome", because their facies were reminiscent of the Kabuki actor's make-up. ("Kabuki' is a traditional Japanese theatre and means, when directly translated into English, "singing, dancing and performing".)

Niikawa has been Professor and Chairman of Human Genetics at Nagasaki University School of Medicine since 1984 and in 1990 he was appointed as an Honorary Professor of Hunan Medical University, China. In Nagasaki, his research has been directed towards molecular genetics and molecular cytogenetics, with special focus upon genomic imprinting.

## Reference

Niikawa N, Matsuura N, Fukushima Y, Ohsawa T, Kajii T (1981) Kabuki make-up syndrome: a syndrome of mental retardation, unusual facies, large and protruding ears, and postnatal growth deficiency. J Pediatr 99(4): 565–569

# Pallister, P.D.

Pallister-Hall syndrome comprises congenital hypothalamic hamartoblastoma, hypopituitarism, imperforate anus and post axial polydactyly. Inheritance is probably autosomal dominant.

Philip D. Pallister was born in 1920. He graduated from the University of Minnesota in 1944, spent 2 years in the US Army medical corps and then became a general practitioner in Boulder, Montana. He passed the next 35 years in practice in this town, where he was also clinical director at the Montana State Training School, an institution for the mentally retarded. In this latter capacity he developed programmes for medical and surgical rehabilitation, neurological assessment and the management of epilepsy. He also initiated a clinical genetic service, with cytogenetic and biochemical genetic support.

Pallister's long-standing academic interest in mental retardation has led to the publication of 35 articles, including several in which newly recognised conditions were delineated; his name has been used eponymously for a number of these disorders, including the Pallister mosaic syndrome (Pallister-Killian), the Pallister W syndrome and the Pallister ulnar mammary syndrome. He has held clinical faculty appointments at the Universities of Washington and Utah and he received an honorary DSc degree from Montana State University. Pallister's activities have brought him into contact with many prominent medical geneticists, including John Opitz, who was largely responsible for his eponymous immortality, and Jurgen Herrmann, who became his son-in-law.

After his retirement from his private and hospital practice in 1981, Pallister established the Shodair genetics programme in Helena, Montana, which was subsequently directed by Dr John Opitz. In 1995 he was living in Boulder, with his wife, Blanche Willa, to whom he has been married for 53 years. The couple had 15 children (14 boys and one girl). There are also 27 grandchildren and 5 great grandchildren. In their spare moments the Pallisters are occupied with cattle ranching, hunting, fishing, automechanics, photography, computers and, especially, archaeology.

## Reference

Hall JG, Pallister PD, Clarren SK, Beckwith JB, Wiglesworth FW, Fraser FC, Cho S, Benke PJ, Reed SD (1980) Congenital hypothalamic hamartoblastoma, hypopituitarism, imperforate anus and postaxial polydactyly – a new syndrome? Part 1: Clinical, causal and pathogenetic considerations. Am J Med Genet 7: 47–74

# Perlman, M.

Perlman syndrome comprises fetal macrosomia, renal abnormalities, a typical facies and a wide variety of visceral anomalies. Inheritance is probably autosomal recessive.

Max Perlman graduated from Sydney University Medical School in 1961 and undertook his internships and paediatric residency training in Australia, the UK and Israel. From 1970 to 1973 he was the director of the neonatal unit at the Beer Sheva Hospital in Israel and during this period he documented a consanguineous family in which several offspring had been born with the condition which now bears his name.

From 1974 to 1981 Perlman was the neonatologist-in-charge of the Neonatal Unit at the Hadassah Hospital in Ein Karem, Jerusalem, Israel. Since 1981 he has been a staff neonatologist at the Hospital for Sick Children, Toronto. Perlman stated "In research I have moved from field to field, attracted by clinical opportunities as they were presented to me. My present focus on risk prediction is leading me into epidemiological questions related to causation and prognosis of disease. I believe that clinicians, by observing experiments of nature and patient outcomes, can still stimulate bedside to bench initiatives as well as find answers to clinical questions."

## Reference

Perlman M, Goldberg GM, Bar-Ziv J, Danovitch G (1973) Renal hamartomas and nephroblastomatosis with fetal gigantism: a familial syndrome. J Pediatr 83: 414–418

# Psaume, J.

The Papillon-Léage and Psaume syndrome, otherwise known as oro-facial-digital syndrome type I, comprises clefts of the tongue and palate, abnormalities of dentition and digital malformations. The mode of inheritance is probably X-linked dominant with male lethality.

Jean Psaume was born in Paris in 1920. His father, Marcel Psaume (1889–1958) was a member of the medical profession and his son Jean followed a similar career. After qualifiying in medicine Jean Psaume obtained an appointment as assistant to Professor Pierre Petit at the Hospital St Vincent de Paul, Paris. Petit was an expert in the surgical closure of facial clefts, and in 1950 Psaume obtained a doctorate with a thesis on this subject. Three years later, he analysed 500 sets of case notes pertaining to patients treated by Petit and identified 8 persons with facial, dental and digital malformations. In 1954, he documented these findings and as his senior colleague, Madame Papillon-Léage[1] had been associated with his investigation, her name appeared on the publication. The conjoined eponym, Papillon-Léage and Psaume syndrome then came into use, although it would also have been appropriate for Petit's name to have been attached to the condition. In 1962, Psaume published an article on the disorder in conjunction with Gorlin, in which the descriptive term "oro-facial-digitial syndrome type I" was employed; thereafter this format gained general acceptance.

During the period 1950–1994 Psaume wrote more than 30 articles on facial malformations, several of which contained details of his pioneering work in the field of surgical management. He also wrote a book on the treatment of cleft lip, which he co-authored with Petit (1961). In the later stages of his career, Psaume occupied a senior post at the Saint Michael Hospital, Paris and following his retirement, he retained honorary status. In 1994 Psaume was living in the peaceful town of Port Grimaud where he was able to pursue his interest in sailing.

## References


Gorlin RJ, Psaume J (1962) Orodigitofacial dysostosis – a new syndrome? J Pediatr 61: 520–530

Papillon-Léage Mme, Psaume J (1954) Une malformation héréditaire de la muqueuse buccale brides et freins anormaux. Rev Stomat (Paris) 55: 209–227


---

[1]Mme Papillon-Léage was one of the first women to specialise in oro-facial surgery. Her name is sometimes confused with that of E. Papillon, a French dematologist who collaborated with P. E. Lefèvre in the delineation of hyperkeratosis palmoplantaris.

# Rallison, M.L.

Wolcott-Rallison syndrome (*see* Wolcott, p. 197).

Marvin L. Rallison was born in Coalville, Utah, USA on 8 February 1929. After studying chemistry at Utah State University, Logan, he qualified in medicine in 1957 at the University of Utah in Salt Lake City. Rallison then married Beth West of Idaho Falls, Idaho and subsequently trained in paediatrics and endocrinology at the University of Minnesota. He currently holds the post of Chief of Pediatric Endocrinology at the University of Utah School of Medicine.

In the late 1960s Rallison participated in an epidemiologic study of thyroid disease in young people who lived near the Nevada atomic test site. This investigation led to original observations on the natural history of adolescent goitre, autoimmune thyroiditis and nodular thyroid disorders in youth. His interest in diabetes began during his fellowship in endocrinology, with medical supervision of a camp for diabetic children and continued with involvement in patient education and management. These contributions were recognised with an award by the American Diabetes Association. He has published a number of articles in this field and written a reference text entitled: *Growth Problems in Infancy, Childhood and Adolescence*. In 1972, in collaboration with his student, Carol Wolcott, Rallison delineated the syndrome of insulin dependent diabetes and multiple epiphyseal dysplasia, which now bears their conjoined eponym.

Rallison stated "When not involved in teaching endocrinology in class and clinic and in supervising research in diabetes and growth problems, I spend time with my wife, our four children (two MD's, a lawyer, and a dietician), and a growing pod of exceptional grandchildren. I potter about in our half-acre yard, renew my love of nature at our cabin in the woods and seek good trout streams. I play duffer golf and tennis and watch birds at every opportunity; ornithology is the real outdoor passion of my life, and it has led me and my family into many byways."

## Reference


Wolcott CD, Rallison ML (1972) Infancy-onset diabetes mellitus and multiple epiphyseal dysplasia. J Pediatr 80(2): 292–297


# Riley, H.D. Jr.

Riley-Smith syndrome comprises macrocephaly, pseudopapilloedema and multiple haemangiomata. Inheritance is probably autosomal dominant.

Harris D. Riley, Jr., was born in Clarksdale, Mississippi and grew up in Tupelo, Mississippi. He attended Vanderbilt University, received a BA degree and proceeded to the School of Medicine where he obtained his MD in 1948. After internships at the Baltimore City and Johns Hopkins Hospitals, Riley completed residency training in paediatrics at the Babies and Children's Hospital, Case Western Reserve University and at Vanderbilt University Hospital. Thereafter he undertook fellowship training in paediatric infectious diseases at Vanderbilt.

Riley served in the US Navy in World War II and in the US Air Force medical service during the Korean War. In the latter he had appointments as head of the paediatric service and chief of the infectious disease division of several hospitals. Riley was an instructor in paediatrics at Vanderbilt University School of Medicine from 1953 to 1957. In 1958 he became professor of paediatrics and head of the department of paediatrics, University of Oklahoma College of Medicine. He was also medical director and chief of staff of the Children's Hospital of Oklahoma, University of Oklahoma Health Sciences Center and chief of the infectious disease division. In 1976 Riley was appointed as distinguished professor of paediatrics at the University of Oklahoma, being the first faculty member on the Health Sciences Center campus to be accorded this honour. He was also professor of human ecology in the College of Health, University of Oklahoma. Riley is married and has 3 children. In 1991 he was appointed professor of paediatrics at Vanderbilt University School of Medicine.

Riley is a member of many scientific societies, including the Society for Pediatric Research, the American Pediatric Society, the Infectious Disease Society of America and the American Academy of Pediatrics. He has been a consultant to various organisations, including the National Institutes of Health, the Center for Disease Control, the US Army, the US Air Force Medical Services and other agencies and foundations. He is a member of the editorial board of several medical journals and has edited the *Southern Medical Journal*.

Riley has been honoured by the establishment of the Harris D. Riley, Jr. Pediatric Society at the University of Oklahoma Health Sciences Center. This society was established by former house officers, fellows and faculty who trained under or were associated with Riley at the University of Oklahoma. Among other activities, the Society sponsors an annual Riley lectureship in paediatric infectious diseases at the University of Oklahoma. The infectious disease unit at the Children's Hospital of Oklahoma is also named in Riley's honour.

## Reference


Riley HD Jr, Smith WR (1960) Macrocephaly, pseudopapilledema and multiple hemangiomata. A previously undescribed heredofamilial syndrome. Pediatrics 26: 293–300

# Rimoin, D.L.

Rimoin syndrome is an autosomal recessive disorder comprising short limbed dwarfism, metaphyseal dysostosis, conductive hearing loss and mild mental retardation.

David Lawrence Rimoin was born in Montreal, Quebec, Canada on 9 November 1936. He attended McGill University where he obtained an MSc degree and a medical qualification in 1961. He then trained in internal medicine and medical genetics at the Royal Victoria Hospital, Montreal and at the Johns Hopkins Hospital, Baltimore, USA, where he proceeded to a doctorate. Rimoin was appointed as head of the genetic clinic at the Washington University School of Medicine, St Louis, in 1967 and 3 years later he moved to the University of California Medical Center, Torrance. In 1973 Rimoin was promoted to professor of paediatrics and medicine, with headship of the division of medical genetics and consultant status at the Cedars-Mt Sinai Hospital, Los Angeles.

Rimoin has been a highly productive researcher and medical author in several fields, including dysmorphology, endocrinology and bone dysplasias. He has co-edited several books, notably the comprehensive *Principles and Practice of Medical Genetics* and he has served as associate editor of a number of genetic and endocrinological journals.

## Reference

Rimoin DL, McAlister WH (1971) Metaphyseal dysostosis, conductive hearing loss and mental retardation: a recessively inherited syndrome. Birth Defects: Orig Art Ser VII/4: 116–122

# Ruvalcaba, R.H.A.

Ruvalcaba syndrome is an autosomal dominant disorder in which microcephaly and an unusual facies are associated with shortened bones in the extremities and changes in the spine.

Rogelio H. Ruvalcaba was born in 1934 in a rural Mexican community, during the great depression. His ambitions for a career in medicine became apparent during boyhood and he subsequently stated "despite the intermittent folk medicine he and his seven younger siblings experienced…painful eucalyptus oil injections for the flu and banana peel applied to the soles of their feet for sore throats…he remained fascinated by medicine and was determined to become a physician". Ruvalcaba received his MD degree from the Universidad de Guadalajara, Mexico at the age of 23 years. His early professional experiences included data gathering in the first trials of chloramphenicol during a salmonella epidemic, service in a leprosarium and a hospital for tuberculosis and tropical disease, and a prevaccine poliomyelitis epidemic in northern Mexico.

Ruvalcaba emigrated to the United States in 1958 and he undertook a second internship and first-year paediatric residency at the Hôtel Dieu Hospital, Louisiana State University in New Orleans. He completed the last 2 years of paediatric residency in Omaha, Nebraska, where he met and married Elaine T. Hermanson, a registered nurse. The Ruvalcabas then moved to Seattle, Washington, where their 4 children, Roger, Randal, Rachelle and Ralaina were born. At the University of Washington he was a fellow in paediatric endocrinology, becoming board certified in both paediatrics and paediatric endocrinology. In 1965, Ruvalcaba became a member of the University of Washington faculty where he currently holds the rank of Professor. His hobbies include photography, gardening, painting, aerobics and music appreciation.

## Reference

Ruvalcaba RHA, Reichert A, Smith DW (1971) A new familial syndrome with osseous dysplasia and mental deficiency. J Pediatr 79: 450–455

# Sanfilippo, S.J.

Sanfilippo Syndrome or mucopolysaccharidosis type III is an heterogeneous autosomal recessive metabolic disorder characterised by progressive mental deterioration and excessive urinary excretion of heparitin sulphate.

Sylvester Sanfilippo was born in Rochester, New York, USA on 1 January 1926. After graduating from the University of Rochester, he moved to Salt Lake City to pursue postgraduate studies at the University of Utah. There he received a master of science degree in biochemistry and earned his medical qualification in 1955.

Sanfilippo trained in paediatrics at the University of Minnesota, interrupted by a 2-year stint as a paediatrician in the United States Navy Medical Corps in Norfolk, Virginia. He was awarded a postdoctoral research fellowship in 1960 and began a comprehensive study of children with mucopolysaccharide storage disease. The investigative approach combined the chemical measurement and identification of urinary acid mucopolysaccharides with a clinical evaluation of each patient. Together with his colleagues Sanfilippo identified 8 mentally retarded children with mucopolysacchariduria of a single compound, heparitin sulphate. The majority of these children had a normal or near-normal facial appearance and displayed mild to slight somatic and radiographic manifestations. The results of the study were presented at the annual American Pediatric Society meeting in May 1963 and published later that year.

Sanfilippo entered private practice in paediatrics in April 1962 but continued his research at the University of Minnesota for several more years. He also enjoyed teaching experiences with medical students and physicians-in-training. Sanfilippo was involved in health care planning and in 1976 he published a perinatal mortality review study. In addition he was an elected officer of local paediatric societies and hospital committees. Sanfilippo retired from private practice in June 1988 and in 1995 he was living in Edina, Minneapolis.

## Reference

Sanfilippo SJ, Podosin R, Langer L, Good RA (1963) Mental retardation associated with acid mucopolysacchariduria (heparitin sulfate type). J Pediatr 63: 837–838

# Schinzel, A.A.G.L.

Schinzel acrocallosal syndrome comprises mental retardation, an unusual facies, polydactyly and absence of the corpus callosum. Inheritance is autosomal recessive.

Albert A.G.L. Schinzel was born in Vienna, Austria on 13 September 1944. His family moved to Innsbruck in 1954 when his father was appointed professor of microbiology at the University in that city. After completing his schooling Schinzel studied medicine at the Universities of Vienna, West Berlin and Innsbruck, qualifying in 1968. Following his internship, he became a research fellow in medical genetics in the department of paediatrics at the University of Zurich. He received steady promotion, receiving full professorial status in 1986.

Schinzel's research interest have included cytogenetics, dysmorphology, teratology and molecular cytogenetics. He has published numerous articles and has served on the editorial boards of several genetic journals. In addition to his extensive clinical commitments, Schinzel participates in teaching programmes and is a member of several international committees.

He is married and has 3 children.

## Reference

Schinzel A (1979) Postaxial polydactyly, hallux duplication, absence of the corpus callosum, macroencephaly and severe mental retardation: a new syndrome. Helv Paediatr Acta 34: 141–146

# Schmid, F.

Schmid metaphyseal chondrodysplasia is an autosomal dominant disorder which presents with stunted stature and leg bowing.

Franz Schmid was born in Lauterbach on 13 March 1920, and educated in that town and in Graslitz-Erzgebirge. He entered the Karls University medical school, Prague in 1938. His studies were interrupted between 1940 and 1942 by military service in World War II, and he subsequently attended the universities of Konigsberg, Breslau and Jena, receiving his medical qualification in 1945.

Schmid obtained wide postgraduate experience at the Children's Hospital, University of Heidelberg before specialising in paediatrics. He became head of the radiology department in 1948 and remained in this post until 1967 when he was appointed as chief physician at the Municipal Children's Hospital, Aschaffenburg. He also became medical director of the Aschaffenburg Municipal Hospital during this period.

Schmid's extensive publications include more than 80 monographs and chapters and over 760 medical articles. He has also written poetry and contributed to the non-medical literature. In the later stages of his career Schmid became interested in cell biology and he was associated with a number of journals devoted to that subject. His contributions in this field received recognition in 1981 when he was elected to the chairmanship of the International Research Association for Cell Therapy.

## Reference

Schmid F (1949) Beitrag zur Dysostosis enchondralis metaphysaria. Mschr Kinderheilk 97: 393–397

# Schmid, W.

Schmid-Fraccaro syndrome, also known as cat eye syndrome, comprises colobomata of the iris and choroid, imperforate anus, pre-auricular skin tags and malformation of the cardiovascular and urinary systems. A supernumerary acrocentric chromosome is present in some affected persons.

Werner Schmid was born in Winterthur, Switzerland on 13 February 1930. He received his schooling at the Gymnasium in Winterthur and qualified in medicine in 1955, after studies at the Universities of Paris and Zurich. Schmid developed a special interest in genetics at an early stage of his career and in 1958 he received a doctorate for a thesis entitled *X-linked recessive visible mutations in Drosophila melanogaster*. He then undertook residence training in pathology in Zurich and in 1960 he became a research associate in cytogenetics at the University of Texas, USA. Schmid returned to the University of Zurich in 1963 and thereafter was appointed as head of a research laboratory and of the division of medical genetics, within the department of paediatrics. In 1978 he became full professor and director of the University Institute of Medical Genetics. Schmid's academic interests have been focussed on cytogenetics and mutation, and he has been responsible for the development of several novel laboratory techniques. He is the author of more than 160 publications and has the distinction of founding the Swiss Society of Medical Genetics (*see The Man Behind the Syndrome* Fraccaro, p. 211).

## Reference

Schachenmann G, Schmid W, Fraccaro M, Mannini A, Tiepolo L, Perona GP, Sartori E (1965) Chromosomes in coloboma and anal atresia. Lancet II: 290

# Sipple, J.H.

Sipple syndrome, or multiple endocrine neoplasia II, comprises medullar thyroid carcinoma, pheochromocytoma and parathyroid adenoma. Inheritance is autosomal dominant.

John H. Sipple was born on 1 July 1930 in Lakewood, Ohio, USA. He graduated from Cornell University and Cornell University Medical College and undertook residency training in internal medicine at the State University of New York Medical Center in Syracuse from 1955 to 1959. During his last year of residency he cared for a hypertensive man who was found to have bilateral pheochromocytomas, bilateral carcinoma of the thyroid gland (later determined to be medullary carcinoma) and a parathyroid adenoma. Sipple found reports of 5 similar cases in the literature (all reported for other reasons) and published an account of the disorder in 1961. Within a few years it became apparent that the syndrome was inherited as an autosomal dominant trait. It was classified as multiple endocrine neoplasia, type 2A, and given the eponym Sipple Syndrome.

Following a fellowship in pulmonary disease at the Johns Hopkins Hospital, Baltimore, Sipple returned to Syracuse in 1962 to practice in pulmonary disease and internal medicine. He has been clinical professor of medicine at the State University of New York Health Science Center in Syracuse since 1977. From 1989 to 1993 Sipple was governor of the Upstate New York Region of the American College of Physicians and in 1995 he became president of the Internist Associates of Central New York.

## Reference

Sipple J (1961) The association of pheochromocytoma with carcinoma of the thyroid gland. Am J Med 31: 163–166

# Sjögren, H.S.C.

Sjögren syndrome comprises keratoconjunctivitis, xerostomia and a variable arthropathy. Familial clustering and HLA associations have been demonstrated, but the genetic basis is uncertain.

Henrik Samuel Conrad Sjögren was born on 23 October 1899 in Köping, Scheele, Sweden. He qualified in medicine at the Karolinska Institute, Stockholm in 1927 and thereafter married Maria Hellgren, the daughter of a prominent ophthalmologist. After completing his specialist training in ophthalmology at the Serafimerlasarettet, he took up an appointment at the Sabbatsberg City Hospital, Stockholm. Sjögren had previously examined a middle-aged woman with the condition which now bears his name, and in 1933 he produced a doctoral thesis on this topic, in which he presented details of 19 affected persons. His thesis was not of a sufficiently high standard for the award of the title of "docent" and thus denied the opportunity for a career in academic ophthalmology, he took up an appointment at the Jönköping City Hospital, Southern Sweden. Following the general acceptance of the syndrome which he had delineated, Sjögren received academic recognition and in 1957 he was made a docent of the University of Gothenburg; 3 years later, he received the title of honorary professor. Sjögren remained at Jönköping until his retirement in 1967 and thereafter, he lived in Lund. In May 1986, an International Seminar on the Sjögren syndrome was held in Copenhagen. Sjögren was to have been the guest of honour, but ill health precluded his participation. He died in a nursing home in Ribbingska, Lund, on 17 September 1986.

## Reference

Sjögren HS (1933) Zur Kenntnis der Keratoconjunctivitis sicca (Keratitis filiformis bei Hypofunktion der Tränendrüsen). Acta Ophth, Copenhagen Suppl 2: 1–151

# Sjögren, T.

Sjögren-Larsson[1] syndrome comprises ichthyosis, spastic paraplegia and mental retardation. Inheritance is autosomal recessive.

Torsten Sjögren was born in Sweden in 1896. He qualified in medicine and trained in psychiatry, obtaining an appointment in the University Clinic in Lund. He subsequently became the first medical administrator at the Sahlgrenska Hospital in Gothenburg where he was instrumental in promoting the establishment of a psychiatric unit. In 1935, Sjögren was appointed as head psychiatrist at this hospital and he was able to make important contributions in the fields of patient care, education and research. A decade later he became professor of psychiatry at the Karolinska hospital. In this phase of his career, he became interested in the genetic aspects of mental retardation and in 1957, in collaboration with Larsson[1], he documented 28 persons with the condition which now bears their names. Sjögren was also involved in the development of genetic counselling facilities, the training of researchers and the implementation of new forms of therapy.

Sjögren is remembered as one of the pioneers of modern Swedish psychiatry. He died on 27 July 1974, survived by his wife, but the couple had no children.

## Reference

Sjögren T, Larsson T (1957) Oligophrenia in combination with congenital ichthyosis and spastic disorders. A clinical and genetic study. Acta psychiatr Scand 32(Suppl 113): 1–112

---

[1]Tage Larsson, born 1905 was a medical genetic statistician at the Karolinska Institute, Sweden.

# Soulier, J.-P.

Bernard-Soulier syndrome (*see* Bernard, p. 206).

Jean-Pierre Soulier was born on 14 September 1915. He qualified in medicine in Paris in 1937 and thereafter undertook specialist training. In 1945 Soulier spent a year at Harvard as a research fellow and upon his return to Paris he became laboratory head at the National Blood Transfusion Service. He acquired postgraduate qualifications and in 1954 he was appointed as director of the National Blood Transfusion Service, where he remained until his retirement in 1984. In 1961 Soulier was elevated to the status of professor of haematology at the University of Paris, at the Necker Hospital for Sick Children.

During four decades of scientific activity, Soulier pioneered many developments in his field and his work on plasma fractionation has had a profound influence on the management of haemophilia and other bleeding disorders. He was a founder member of the committee for the nomenclature of blood clotting factors, which evolved into the International Society on Thrombosis and Haemostasis. For many years Soulier has served as secretary of the International Society for Blood Transfusion and in 1970 he was president of the French Haematology Society. In 1975 Soulier was elected as president of the International Congress of Thrombosis and Haemostasis and from 1978 to 1980 he was president of the International Society of Blood Transfusion.

In 1994, at the age of 79 years, Soulier was living in retirement in Paris.

## Reference

Bernard J, Soulier J-P (1948) Sur une nouvelle variété de dystrophie thrombocytaire-hémorragipare congénitale. Sem Hôp Paris 24: 3217–3223

# Steinert, H.

Steinert syndrome or dystrophia myotonica is characterised by muscle stiffness, cataracts, hypogonadism and low intelligence. Inheritance is autosomal dominant, with very variable clinical expression.

Hans Steinert was born in Freiberg on 10 April 1875 and received his schooling at the Gymnasium in that city. He subsequently studied medicine at the Universities of Leipzig, Freiberg, Berlin and Kiel and graduated in 1898. After junior appointments in Halle and Berlin, he moved to Leipzig where he worked firstly in the Pathological Institute, then in the Medical Clinic and achieved professorial status in 1910.

Steinert was a productive medical author, mainly in the field of neurological and muscular conditions. Shortly before his appointment at Leipzig he described the muscle disorder which is now known as dystrophia myotonica and which sometimes carries his eponym.

## Reference

Steinert H (1909) Myopathologische Beitrage. 1. über das klinische und anatomische Bild des Muskelschwunds der Myotoniker. Deut Zshr Nervenheilk 37: 58–104

# Verma, I.

Verma-Naumoff syndrome is an autosomal recessive, lethal, short-rib dwarfing skeletal dysplasia.

Ishwar Verma was born on 25 December 1936 in Eldoret, Kenya, where his father was a railway engineer. He received his early schooling in Eldoret and qualified in medicine at the University of Punjab in Amritsar, India. Verma then returned to Africa in order to work in the university hospital in Dar-es-Salaam, Tanzania. He subsequently trained in medicine and paediatrics in the United Kingdom and, after obtaining a higher qualification, moved in 1967 to the All India Institute of Medical Sciences, New Delhi. Verma acquired additional experience in medical genetics in Zurich and at the Massachusetts General Hospital, Boston and in 1988 he was appointed as professor of paediatrics and director of the genetics unit at the All India Institute.

Verma has made significant academic contributions concerning genetic disorders and congenital malformations in India. In particular, he has documented medicogenetic problems of tribal communities and been involved in the molecular characterisation of beta-thalassaemia and Duchenne muscular dystrophy. In addition to his academic activities, Verma has established extensive genetic services for counselling and prenatal diagnosis; his genetic unit is pre-eminent in India, and is recognised as a WHO collaborating centre. His pioneering work in the field of medical genetics in India has been recognised by the award of numerous honours and distinctions.

Verma's family has a strong medical tradition; his brother and 2 sisters are physicians, his wife is a cyto-pathologist and his son has qualified in medicine.

## Reference

Verma IC, Bhargava S, Agarwal S (1975) An autosomal recessive form of lethal chondrodystrophy with severe thoracic narrowing, rhizoacromelic type of micromelia, polydactyly and genital anomalies. Birth Defects Orig Art Ser XI(6): 167–174

# Ward, O.C.

[1]Romano-Ward syndrome manifests as episodic cardiac arrhythmia, which leads to syncope or sudden death. The electrocardiogram shows a characteristic prolonged Q-T interval.

O. Conor Ward was born in Monaghan, Ireland, in 1923, the eldest son of Dr F.C. Ward, a general practitioner who was also parliamentary secretary for public health in the Irish government. At the time of his graduation in medicine at the University College, Dublin in 1947, he was awarded the Staunton Memorial gold medal in paediatrics. Ward trained in paediatrics and paediatric cardiology at the University of Liverpool and in 1951 he received a doctorate. He was admitted as a fellow of the Royal College of Physicians of Ireland, becoming censor in 1978 and vice-president in 1979. In 1982 Ward was foundation dean of the college faculty of paediatrics. He became professor of paediatrics of the University College, Dublin in 1972 and was consultant paediatrician and head of paediatric cardiology at Our Lady's Hospital for Sick Children. Ward established a bereavement counselling service for cot death families and an advisory service for Down syndrome patients.

After his retirement he represented the Royal College of Surgeons of Ireland as visiting professor and chief of paediatrics in Tabuk in Saudi Arabia. Ward also served as chairman of the European Association of Paediatric Cardiologists, president of the Irish Paediatric Association, the Irish Cardiac Society, the Irish Heart Foundation and the Association for the Welfare of Children in Hospital. He has been external examiner to universities in the UK, the Middle East and Africa and he has published widely on paediatrics and paediatric cardiology.

In 1994 Ward was living in retirement in London where he held honorary teaching appointments.

## Reference

Ward OC (1964) New familial cardiac syndrome in children. J Irish Med Assoc 322: 103–106

---

[1]C. Romano is an Italian physician.

# Watson, G.H.

Watson syndrome is an autosomal dominant disorder comprising café-au-lait macules, pulmonary stenosis and dull intelligence. The condition probably represents a sub-type of neurofibromatosis.

Geoffrey H. Watson was born in 1920 in Oldham, an industrial town in the north of England. He was the only child of a school headmaster and his wife, who was also a teacher. Watson won open scholarships to the Oldham Grammar School and, later, to Manchester University, where he obtained a BSc degree and subsequently graduated in medicine in 1944.

After periods in the Royal Navy and in junior medical posts, Watson became interested in paediatrics and spent an enjoyable, hardworking year learning about paediatric cardiology from John Keith and Richard Rowe in Toronto. Although Watson and his wife found Canada very attractive they returned to England, where he became a lecturer in the faculty of medicine at the University of Manchester and was appointed as a consultant in paediatrics and paediatric cardiology.

At the Royal Manchester Children's Hospital, Watson established a cardiac diagnostic service, using effective and up-to-date but perforce extremely inexpensive catheterisation and cine-angiocardiography apparatus. He then successfully encouraged the development of open heart surgery for children at the same hospital. Watson remained single-handed in the field of paediatric cardiology until 1979 but he was also able to run a general paediatric service. In 1987 he eased himself into retirement by acting for 2 years as his own part-time research fellow, investigating the cardiac aspects of tuberous sclerosis.

## Reference

Watson GH (1967) Pulmonary stenosis, café-au-lait spots and dull intelligence. Arch Dis Childh 42: 303–307

# Weaver, D.D.

Weaver syndrome comprises overgrowth, advanced skeletal maturation and an unusual facies. Inheritance is probably autosomal recessive.

David Weaver was born and raised in Twin Falls, Idaho, USA, a farming community, and educated at a small liberal arts institution near Boise, Idaho. He obtained his medical training at the University of Oregon School of Medicine (now the Oregon State Health Sciences Center) and graduated in 1966 with a MSc degree in anatomy and as a doctor of medicine. From 1967 to 1970 Weaver served in the US Public Health Service at the Arctic Health Research Center in Fairbanks, Alaska, where he was involved in biochemical genetic studies of the Alaskan Eskimos and Indians. He then undertook a 2-year residency in paediatrics at the University of Oregon Medical School, followed by a medical genetics fellowship at the University of Washington School of Medicine, Seattle.

During his fellowship, Weaver was greatly influenced by Dr David Smith, who introduced him to dysmorphology, and by Dr Judith Hall who further developed his interest in clinical genetics. In 1974 Weaver returned to the medical school at Portland, Oregon, for a fellowship in metabolism, and since 1976 he has been a member of the faculty of the department of medical and molecular genetics, Indiana University School of Medicine, Indianapolis, where he is director of clinical activities. Weaver's research interests include syndrome delineation and mechanisms which produce birth defects.

## Reference

Weaver DD, Graham CB, Thomas IT, Smith DW (1974) A new overgrowth syndrome with accelerated skeletal maturation, unusual facies and camptodactyly. J Pediatr 84: 547–552

# Wolf, U.

Wolf-Hirschhorn syndrome comprises mental and physical retardation, a characteristic facies and multiple developmental abnormalities. The condition results from a deletion of part of the short arm of chromosome 4.

[1]Ulrich Wolf was born in 1933 in Riesa, Saxonia, Germany and he spent his youth in Leipzig. He studied biology at the Universities of Tübingen and Munich, obtaining a doctorate in biology in 1961. He then trained in cytogenetics with Klaus Pätau at Madison, Wisconsin and in human genetics with Helmut Baitsch at Freiburg, Germany. In 1972 he was appointed to his current post as professor of human genetics at the University of Freiburg.

Wolf is a member of the Deutsche Akademie der Naturforscher Leopoldina and a Fellow of the Institute for Advanced Study, Berlin. In 1975, he was elected as president of the European Society of Human Genetics. In addition to his university activities, Wolf is the editor of the journal *Human Genetics*. His current scientific interests are the mechanisms of sex determination and genotype-phenotype relationships. Outside his academic life Wolf has a keen interest in the philosophy of science and he also enjoys playing the violincello and contrabass.

(*see* Hirschhorn, p. 212.)

## References

Wolf U, Porsch R, Baitsch H, Reinwein H (1965) Deletion on short arms of a B-chromosome without "cri-du-chat" syndrome. Lancet, April 3: 769

Wolf U, Reinwein H, Porsch R, Schröter R, Baitsch H (1965) Defizienz an den kurzen Armen eines Chromosoms Nr. 4. Humangenetik 1: 397–413

---

[1]Louis Wolff, of the Wolff-Parkinson-White syndrome, was a cardiologist in the USA. (See p. 193)

# Appendix

The subjects of *The Man Behind the Syndrome* were as follows:

## Section I.  Portraits and Biographies

ALBERS-SCHÖNBERG, Heinrich Ernst *(1865–1921)*
ALBRIGHT, Fuller *(1900–1969)*
ALPORT, Arthur Cecil *(1880–1959)*
ALZHEIMER, Alois *(1864–1915)*
APERT, Eugene *(1868–1940)*
BATTEN, Frederick Eustace *(1866–1918)*
BELL, Julia *(1879–1979)*
BIEDL, Arthur *(1869–1933)*
BLACKFAN, Kenneth Daniel *(1883–1941)*
BRAILSFORD, James Frederick *(1888–1961)*
CAFFEY, John Patrick *(1895–1978)*
CARPENTER, George Alfred *(1859–1910)*
CHARCOT, Jean Martin *(1825–1893)*
COCKAYNE, Edward Alfred *(1880–1956)*
CREUTZFELDT, Hans-Gerhard *(1885–1964)*
CROUZON, Octave *(1874–1938)*
DANLOS, Henri-Alexandre *(1844–1912)*
DEJERINE, Joseph Jules *(1849–1917)*
DE LANGE, Cornelia *(1871–1950)*
DOWN, John Langdon Haydon *(1828–1896)*
DUANE, Alexander *(1858–1926)*
DUCHENNE, Guillaume Benjamin Amand *(1806–1875)*
DUPUYTREN, Guillaume *(1777–1835)*
EHLERS, Edvard *(1863–1937)*
ELLIS, Richard White Bernard *(1902–1966)*
FABRY, Johannes *(1860–1930)*
FAIRBANK, Harold Arthur Thomas *(1876–1961)*
FANCONI, Guido *(1892–1979)*
FRANCESCHETTI, Adolphe *(1896–1968)*
FRANÇOIS, Jules *(1907–1984)*
FREEMAN, Ernest Arthur *(1900–1975)*
FRIEDREICH, Nikolaus *(1825–1882)*
GAUCHER, Phillipe Charles Ernest *(1854–1918)*
GILBERT, Nicolas Augustin *(1858–1927)*
GREIG, David Middleton *(1864–1936)*
HEBERDEN, William *(1710–1801)*
HIRSCHSPRUNG, Harald *(1830–1916)*
HOFFMANN, Johann *(1857–1919)*
HUNTER, Charles *(1873–1955)*
HUNTINGTON, George Sumner *(1850–1916)*
HURLER, Gertrud *(1889–1965)*
JAFFE, Henry *(1896–1979)*
JAKOB, Alfons Maria *(1884–1931)*
JANSEN, Murk *(1867–1935)*
KARTAGENER, Manes *(1897–1975)*
KLEIN, David *(1908–    )*
KRABBE, Knud *(1885–1965)*
LANDOUZY, Louis Théophile Joseph *(1845–1917)*

LAURENCE, John Zachariah *(1829–1870)*
LEBER, Theodor *(1840–1917)*
LICHTENSTEIN, Louis *(1906–1977)*
MADELUNG, Otto Wilhelm *(1846–1926)*
MARFAN, Bernard Jean Antonin *(1858–1942)*
MARIE, Pierre *(1853–1940)*
MECKEL, Johann Friedrich the Younger *(1781–1833)*
MENIÈRE, Prosper *(1799–1862)*
MILROY, William Forsyth *(1855–1942)*
MOEBIUS, Paul Julius *(1853–1907)*
MOHR, Otto Louis *(1886–1967)*
MOON, Robert Charles *(1845–1914)*
MORQUIO, Luis *(1867–1935)*
NORRIE, Gordon *(1855–1941)*
OLLIER, Louis Xavier Edouard Leopold *(1830–1900)*
OPPENHEIM, Hermann *(1858–1919)*
OSLER, William *(1849–1919)*
PAGET, James *(1814–1899)*
PEUTZ, Johannes Laurentius Augustinus *(1886–1957)*
PICK, Arnold *(1851–1924)*
PICK, Ludwig *(1868–1944)*
POMPE, Johannes Cassianus *(1901–1945)*
REFSUM, Sigvald *(1907–)*
RENDU, Henri Jules Louis Marie *(1844–1902)*
RIEGER, Herwigh *(1898–)*
ROBERTS, John Bingham *(1852–1924)*
ROTHMUND, August von *(1830–1906)*
ROUSSY, Gustave *(1874–1948)*
SACHS, Bernard *(1858–1944)*
SCHEUERMANN, Holger Werfel *(1877–1960)*
SCHILDER, Paul Ferdinand *(1886–1940)*
SHELDON, Joseph Harold *(1893–1972)*
SMITH, David W. *(1926–1981)*
TAY, Waren *(1843–1927)*
THOMSON, Matthew Sydney *(1894–1969)*
TOOTH, Howard Henry *(1856–1925)*
TOURETTE, GILLES de la, Georges Albert Edouard Brutus
  *(1855–1904)*
TREACHER COLLINS, Edward *(1862–1932)*
TURNER, Henry Hubert *(1892–1970)*
USHER, Charles Howard *(1865–1942)*
VAN BUCHEM, Franz Simon Peter *(1898–1979)*
VAN CREVELD, Simon *(1894–1971)*
VON GIERKE, Edgar Otto Konrad *(1877–1945)*
VON RECKLINGHAUSEN, Friedrich Daniel *(1833–1910)*
VON WILLEBRAND, Erik Adolf *(1870–1949)*
WAARDENBURG, Petrus Johannes *(1886–1979)*
WEBER, Frederick Parkes *(1863–1962)*
WERDNIG, Guido *(1844–1919)*
WERNER, Carl Wilhelm Otto *(1879–1936)*
WIEDEMANN, Hans-Rudolf *(1915–)*
WILMS, Max *(1867–1918)*
WILSON, Samuel Alexander Kinnier *(1878–1937)*

# Section II.  Brief Biographies

AARSKOG, Dagfinn
ALSTRÖM, Carl Henry
AUSTIN, James H.
BARTTER, Frederic C.
BECKER, Peter Emil
BECKWITH, J. Bruce
BEHR, Carl
BLOUNT, Walter Putnam
CAMPAILLA, Ettore
CAMURATI, Mario
CLAUSEN, Jørgen
COFFIN, Grange Simons
CRIGLER, John Fielding, Jnr
DERCUM, Francis Xavier
DIAMOND, Louis K.
DREIFUSS, Fritz E.
DUBIN, I. Nathan
DUBOWITZ, Victor
DYGGVE, Holger V.
EDWARDS, John Hilton
ELLISON, Edwin H.
EMERY, Alan E.H.
ENGELMANN, Guido
FARBER, Sidney
FRACCARO, Marco
FRASER, George R.
GARDNER, Eldon
GIEDION, Andreas
GOLTZ, Robert W.
GOODMAN, Richard M.
GORLIN, Robert James
HAJDU, Nicholas
HALLERMANN, Wilhelm
HANHART, Ernst
HOLT, Mary
JAMPEL, Robert S.
JARCHO, Saul
JEGHERS, Harold Jos
JOHNSON, Frank B.
KLINEFELTER, Harry Fitch, Jnr
KLIPPEL, Maurice
KOZLOWSKI, Kazimierz
KUGELBERG, Eric
LAMY, Maurice
LANGER, Leonard O.
LARON, Zvi
LARSEN, Loren J.
LENZ, Widukind
LÉRI, André
LÉVY, Gabrielle
LIEBENBERG, Freddie
LINDAU, Arvid
LOWRY, Brian
MAJEWSKI, Frank

MARINESCO, Georges
MAROTEAUX, Pierre
MARSHALL, Don
McARDLE, Brian
McCORT, James J.
McCUNE, Donovan James
McKUSICK, Victor Almon
MELCHIOR, Johannes Christian
MELNICK, John C.
MENKES, John
NAJJAR, Victor Assad
NIEMANN, Albert
NOACK, Margot
NYHAN, William L.
OPITZ, John M.
ORAM, Samuel
PARKINSON, James
PENA, Sergio D.J.
PFEIFFER, Rudolf Artur
POLAND, Alfred
POTTER, Edith
PRADER, Andrea
PYLE, Edwin
REINHARDT, Kurt
RENPENNING, Hans J.
ROBIN, Pierre
ROBINOW, Meinhard
ROMBERG, Moritz Heinrich
ROYER, Pierre
RUBINSTEIN, Jack
SALDINO, Ronald Michael
SANDHOFF, Konrad.
SCHEIE, Harold G.
SCHIMKE, R. Neil
SEIP, Martin Fredrik
SHOKEIR, Mohamed H.K.
SHWACHMAN, Harry
SHY, George Milton
SILVER, Henry K.
SLY, William
SMITH, Roy C.
SORSBY, Arnold
SPRANGER Jürgen
SPRENGEL, Otto Gerhard Karl
STANESCU, Victor
STICKLER, Gunnar Brynolf
STREIFF, Bernardo
STURGE, William Allen
TAYBI, Hooshang
THOMSEN, Asmus Julius Thomas
TREVOR, David
VON HIPPEL, Eugen
WEISMANN-NETTER, Robert
WISKOTT, Alfred
ZELLWEGER, Hans U.
ZOLLINGER, Robert M.

# Index